To Peter

SOUTH Y

Place-Names

Published by Sigma Leisure - an imprint of Sigma Press, Stobart House, Pontyclerc, Penybanc Road, Ammanford, Carmarthenshire SA18 3HP.

British Library Cataloguing in Publication Data
A CIP record for this book is available from the British Library.

ISBN: 978-1-910758-10-6

Typesetting and Design by: Sigma Press, Ammanford, Carmarthenshire

Cover photographs:
main picture: Roadsign near Midhopestones, Stocksbridge © Anthony Poulton-Smith
top (left to right): Hellaby Hall, Bramley (courtesy of Wikipedia Creative Commons); High Hoyland sign (Anthony Poulton-Smith); Silkstone Common (Anthony Poulton-Smith); Brodsworth Hall (courtesy of Wikipedia Creative Commons)

Photographs: © Anthony Poulton-Smith, unless otherwise stated

Printed by: Akcent Media Ltd.

SOUTH YORKSHIRE
Place-Names

Anthony Poulton-Smith

SIGMA Leisure

Contents

Introduction

For years the history of England was based on the Roman occupation. In recent years we have come to realise the influence of the Empire did not completely rewrite British history, indeed there was already a thriving culture in England well before the birth of Christ. When the Romans left our shores in the 5th century the arrival of the Anglo-Saxons was thought to herald a time of turmoil, yet they brought the culture and language which forms the basis of modern England. Later the arrival of the Norsemen saw their influence and the same is true of our place names, the vast majority of settlement names in Yorkshire are derived from the Saxon/Old English or Old Scandinavian tongues. There are also the topographical features such as rivers and hills still have names given to them by the Celts of the pre-Roman era.

Ostensibly place names are simply descriptions of the location, or of the uses and the people who lived there. In the pages that follow an examination of the origins and meanings of the names in Yorkshire will reveal all. Not only will we see Saxon and Scandinavian settlements, but Celtic rivers and hills, Roman roads and even Norman French landlords all having contributed to the evolution of the names we are otherwise so familiar with.

Not only are the basic names discussed but also districts, hills, streams, fields, roads, lanes, streets and public houses. Road and street names are normally of more recent derivation, named after those who played a significant role in the development of a town or revealing what existed in the village before the developers moved in. The benefactors who provided housing and employment in the eighteenth and nineteenth centuries are often forgotten, yet their names live on in the name found on the sign at the end of the street and often have a story to tell. Pub names are almost a language of their own. Again they are not named arbitrarily but are based on the history of the place and can open a new window on the history of our towns and villages.

Defining place names of all varieties can give an insight into history which would otherwise be ignored or even lost. In the ensuing pages we shall examine 2,000 plus years of Yorkshire. While driving around this area the author was delighted by the unique place names found in the county and so, having already taken a look at, among others, Warwickshire, Oxfordshire, Herefordshire, Nottinghamshire, *Cambridge Place Names*, *County Durham Place Names*, and *Northumberland Place Names* turned here to Yorkshire and in this volume South Yorkshire. This book is the result of the author's long interest in place names which has developed over many years and is the latest in a series which continues to intrigue and surprise.

To all who helped in my research, from the librarians who produced the written word to those who pointed a lost traveller in the right direction, a big thank you.

Place-Names

(in alphabetic order)

A

Adwick le Street

From a Saxon personal name and Old English *wic*, here is one of two places of this name telling of 'the specialised (most often dairy) farm of a man called Adda'. There is no suggestion that the two places refer to the same person, simply they shared the same name. The addition, to distinguish this from its namesake, is from *straet* referring to 'the Roman road'.

Here we also find two local names with Old Scandinavian origins. Both feature the descriptive element *thveit* as the suffix which, as in the norm, follows a specific reference to each place. In Hangthwaite we are told it was 'the woodland clearing of a man called Hagni', while Langthwaite Lane is only surviving reminder of what was 'the long woodland clearing'. Note in the latter the term 'long' can be useful if knowing roughly which neighbour named the place. It must have been named by those living further along the road which bears the name for it appeared, from their perspective 'long'. However had they been living ninety degrees further around the compass the place would have been seen as 'broad' and the name would have reflected this.

Adwick le Street sign

Humber Head Dike is a tributary of the Don. It shares its name with the much larger estuary, where the name is uncertain but may come from British or Celtic *su-mbro* speaking of 'the good river'. However here this can hardly be anything other than a transferred name.

The Foresters shows this was a meeting place or lodge for the Ancient Order of Foresters, a friendly society with branches on both sides of the Atlantic.

Adwick upon Dearne

A second place name from a Saxon personal name and Old English *wic* which refers to 'the specialised (probably dairy) farm of a man called Adda'. Here the addition refers to the River Dearne, a name discussed under its own entry.

Hound Hill Dike is derived from *hund hyll dic* and speaks of 'the ditch at the hill where hounds are kept'. Listed as Mill Flatt in 1843 and derived from *myln flat* or 'the mill at the level piece of the common field', the sole reminder today is found in Mill House. Lousy Busky Lane adds Old Scandinavian *buskr* to Middle English *lousi* and speaks of 'the bush at the place infested by insects'.

Field names begin with Adwick Shroggs, where *scrogge* is added to the basic place name to refer to 'the place of brushwood at Adwick'. Tithe Laithe Close is a reminder of the 'tithe barn' once found here. Pingle comes from *pingel* or 'enclosure'. Ley Field is derived from *laege feld* and speaks of 'the open land allowed to lie fallow'. Wash Close is derived from *waersce* or 'washing place', possibly cloth or maybe sheep, or even both.

The Manvers Arms recalls this is on land owned by Earl Manvers, the coalmine sunk here had the same name.

Anston (North and South)

Two places with a common origin in Old English *ana stan* or 'the solitary stone', one which would have provided a marker. These places are recorded in Domesday as Anestan and Litelanstan, yet by 1297 these had become Northanstan and Suthanstan. The modern additions need no explanation.

Street names include Starch House Road, where starch was indeed produced in the nineteenth century. Harry Crofts is a filed named as 'the smallholding of Harry', this being Old Harry, a reference to the Devil.

Here we see minor names such as Brancliffe Grange, this basic name from *brant clif* or 'the steep bank'. The addition is a comparatively recent addition. Woodsetts is a reminder of rather smaller holding areas for livestock as it describes 'the folds in the woodland'. In Gildingwells we find a name describing a good source of water as 'the gushing spring or stream'.

Ardsley

Listed as Erdeslaia in the twelfth century, this features a Saxon personal name and Old English *leah* telling of 'the woodland clearing of a man called Eored or Eanred'.

Hunningley is a minor place name coming from *ing leah* and referring to 'the woodland clearing associated with a man called Hund'. Measborough Dike points out this channel near the 'fort on the boundary', the name from *gemaeres burh*. Bleach Croft is a reminder of this area being used for bleaching cloth. Stairfoot comes from *staeger* or 'stair' and named from the man-made flight cut into a very steep dip in the land.

Arksey

Recorded as Archeseia in Domesday, this name combines a Scandinavian personal name and Old English *eg* to speak of 'the dry ground in a marsh of a man called Arnkel'.

Bentley is a common place name found throughout the country. As with all common names the meaning is simple, here from Old English *beonet leah* and 'the woodland clearing where bent grass grows'. Here defining the name tells us more than is first apparent. Grasses are highly advanced, at least for a plant, growing from the base and not the tip as other plants. Hence grazers can eat the leaves and the plant suffers no ill effects, indeed grasses actually benefit from being cropped regularly. As successful as grasses are they do not grow bent but will collapse under its own weight eventually, particularly when affected by inclement weather. This tells us the grass was not grazed by livestock but also tells us wild animals did not have access to this area and thus the 'woodland clearing' was fenced off. This suggests the settlement here was not based on farming, although there is no evidence of any alternative.

Almholme comes from Old Scandinavian *almr holmr* and tells of 'the dry land in a marsh where elm trees grow'. The same suffix is seen in the name of Shaftholme or 'the dry land in a marsh where poles are obtained'. Clearly this is where trees were pollarded to produce straight poles used in building. Stockbridge speaks of itself as 'the bridge made of logs' and a particular favourite of the author for, when defined, it describes an image of the place.

One of the most common pub names in the country is the Plough. Here is an invitation to those working the land to patronise the local, once upon a time virtually the whole nation. Interestingly the name is not seen before the sixteenth century.

One problem faced by present-day residents which the Saxons could never have faced is found at the local level-crossing. Regular commuters will be well aware of the length of time the crossing gates bar road traffic from crossing the line, which can be up to an astonishing nine hours in total in any twenty-four hours.

Armthorpe

Listed as Ernulfestorp in the Domesday record of 1086, here is 'the outlying farmstead of a man called Earnwulf or Aernulfr'. This is an Old Scandinavian place name where the personal name is followed by *thorp*.

Armthorpe Shaw adds *sceaga* or 'copse' to the basic place name. Gunhills is from the dialect *ginnel* or 'a narrow passage between buildings' but sometimes used to describe 'a water channel or gutter'. Holme House comes from *holmr* or 'dry land in the marsh'. Rands Lane features *rand* and tells us this could be found at 'the edge or border'. The same elements are seen in Holmewood Rands, where the addition of *wudu rand* gives 'the wood at the border or edge of the dry land in the marsh'.

Cow House Lane may have changed from the original *cu hus layne* but the definition is still self-evident. Both Fore's Drain and Fore's Bridge share an origin in *furh* or 'furrowed'. The Lyngs has not changed since *lyng* spoke of 'the heather'. From *thorn holmr* or 'the dry land in the marsh where thorn bushes are seen'. From *trani mor* comes Tranmoor or 'the boggy land frequented by cranes'. Whiphills is from *hwippe hyll* and refers to 'the hill where young shoots grow'. The Bellrope Acre is a reminder of how this land was set aside to raise money to provide bell rope for the parish church.

Public houses include the Acorn, where there is unlikely to be a message no matter how creative some explanations may be. Here the fruit of the oak tree will have been a simple visual sign while still linking it to the tree, a common pointer to location. Outside the Wheatsheaf we see a sign used by both the Worshipful Company of Bakers and the Brewers' Company. While the latter is the most obvious both could well be relevant as village inn-keepers required a second string to their bow, local beer sales would never be sufficient to keep them in business.

Askern

The earliest record of this name is from the late twelfth century where it appears exactly as it does today. This is a combination of Old Scandinavian *askr* and Old English *aern* giving 'the house near the ash tree'.Unfortunately the number of early records available for Askern Mather, Mather Dike, and Mather Pits are insufficient for us to decide whether this comes from early Modern English mathers meaning 'stinking camomile' or the Middle English mathra or 'madder'. The root of the latter, *Rubia tinctorum*, was used to produce a dye for the cloth known as Red Madder and also to colour a paint. Aside from resembling true chamomile plants, all the former can claim is to have brought the word 'stink' into the English vocabulary.

Askern sign

Askern is known for its spa waters and has been since the eighteenth century when Dr Short's book *Mineral Waters of Yorkshire* mentioned the village as having waters of a most unpleasant odour and taste. This proved an irresistible draw for the

Victorians who arrived on the newly-laid railway to experience the healing powers of the waters.

Aston

A very common name and one most often found with a second distinguishing element. However here this is simple 'the eastern farmstead' from Old English *east tun* and is found in Domesday as Estone.

Locally we see Hardwick, a fairly common Old English name from *heorde wic*. Here both elements ostensibly refer to the same thing. The suffix is properly defined as 'specialised farmstead' but inevitably that speciality is dairy farming, while the first element can still be recognised as meaning 'herd'. Whilst here this is obviously a herd of cattle, it should be noted this collective noun was also used for sheep, 'flock' is a comparatively recent term.

Auckley

In Domesday this is found as Alchelie, a name featuring a Saxon personal name and Old English *leah* teeling us of 'the woodland clearing of a man called Alca or Alha'.

Locally we see the Carrs, this from *kjarr* and a pointer to 'the marsh'. The same element is seen in Hanging Carr, where *hangende kjarr* speaks of 'the marsh at the overhanging land'. Old English *heg feld* comes Hay Field or 'the open land where hay is obtained'. Hurst Plantation featured the element *hyrst* before this area of trees was laid out, this already speaking of 'the wooded hill' but nothing to do with the managed plantation. Marr Flats comes from *mor flat* or 'the moorland with the level meadow land'. Rands Bank is from *rand* or 'edge, border' and is indeed found on the town's northern border. Warren Farm was renamed, probably from a surname, in 1935 and had previously been marked on the map as Poverty Hall, a reminder the land saw little return.

Locals enjoy a drink at the Eagle and Child, a device taken from the coat of arms of the Stanley family, earls of Derby. It is said the image was adopted after their ancestor, Sir Thomas Latham, had an illegitimate son in the fourteenth century. Wanting to make the child his heir he had it placed under a tree, one where an eagle had its nest, and where it was found by himself and his wife as they wandered this estate and adopted as their own. Whether Sir Thomas had a change of heart in later years is unknown but we do know the boy did not inherit. That went to Isobel, his daughter, who married Sir John Stanley and thus uniting the estate, the name and the image of the eagle and child.

Aughton

The South Yorkshire example is the second in the traditional county of Yorkshire, one recorded in Domesday as Actone. Again this is from Old English *ac tun* and describes 'the farmstead by the oak trees'.

Here we find Melton Court, a reminder of the family who were lords of this manor by 1379. William Melton was Archbishop of York when he purchased the land in 1332, also documented as Keeper of the Wardrobe in the Household of Edward II in 1314 and as Treasurer in 1325. Darcy Close remembers the next family to hold this manor, it passing to George, son of John, Lord Darcy as the husband of the late Sir John Melton's daughter and sole heir in 1545. Holderness Drive recalls Lord Conyers Darcy, created Earl of Holderness in 1682.

Verelst Avenue can be traced to one Harry Verelst. Investing in the East India Company, he became firm friends with Clive of India and eventually succeeded him as Governor of Bengal. Having made his fortune he returned to England rather than his family's ancestral Dutch homeland. Here he married the daughter of prosperous merchant Josiah Wordsworth, Ann. Purchasing the Aughton estate in 1773, he and his descendants lived at Aston Hall until 1928.

Austerfield

Listed as Eostrefeld in the early eighth century and as Oustrefeld in Domesday, this is a name from Old English *eowestre feld* referring to 'the open land with a fold for sheep'.

Brancroft comes from *brom croft* and speaks of 'the smallholding where broom grows'. Crow Wood is from *crawe wudu* and is quite clearly 'the wood frequented by crows'. Hag Plantation comes from *hogg* or 'the felling of trees' even before the later managed woodland was planted. Riddings Lane comes from Old English *rydding* to tell us this was 'the clearing'. Southfield House was build at *sud feld* or 'the southern open land'. Willow Holt began as 'the copse where willow grows' from *wilig holt.* Sally Lee features *salh* or 'willow'. Pinfold is from *pynd fald* or 'the pen where stray animals are kept', Pinders Close sharing this derivation.

Dyon's Gate is a reminder of the Diyon, a family first recorded here in the late eighteenth century. Other families have left their mark on the landscape in Holdin Causeway, a reminder of the family of Andrew Holden first seen here in 1636; Hurst House recalls the Hurst family of 1701; in 1768 Jane Spink is recorded at Spink Hall; while Dring Close probably comes from the surname Dring but may also be derived from *drengr* or 'servant'.

B

Barnburgh

Found as Bernebroc in 1086, Barnaburc in 1147, Barneburgh in 1186, as Barmeburgh in 1379, Branburgh in 1507, Baronbrowghe in 1573, and as Barnbroughe in 1638, here is a Saxon personal name with Old English *burh* referring to 'the fortified place of a man called Biarni'.

Locally we see the name of Harlington, an Old English name where *ing tun* follows a personal name and tells of 'the farmstead associated with a man called Herela'. At the Crown Inn we see a pub named to show it was a patriotic establishment.

In the church of St Peter's is the tomb of the Cresacre family, lords of Barnburgh from the thirteenth to the sixteenth centuries. The effigy marking the tomb depicts a knight with a large cat at his feet. This is said to probably refer to Sir Percival Cresacre, a fifteenth century soldier who was returning home through woodland from Doncaster. As he came through the narrow track a wild cat leapt upon the back of his horse, whereupon his steed threw them both to the floor and fled.

Thereafter the cat and the knight fought a long and bloody battle throughout the mile journey until they reached the church of St Peter. However Sir Percival was now fatally wounded and collapsed in the church porch. With his last breath he stretched out and succeeded in crushing the life out of the cat against the stone wall of the porch. To this day the stone slabs are said to be coloured red by the blood spilt by the combatants who were found where they died here by the search party who set out to find them when Sir Percival's horse found his way home.

Barnby Dun

There are two possible Old Scandinavian origins for this name. Despite the difference this name could feature the same elements as Barmby, *barn by*, and thus the same meaning of 'the farmstead of the children'. However it is also possible the first element may be a personal name and then it would be defined as 'the farmstead of a man called Barni or Bjarni'. This name is recorded as Barnebi in 1086 and as Barneby super Don in 1285, the addition distinguishing this from the following entry and describing its location 'on the River Don', a river name discussed under its own entry.

Tilts is an unusual name, it comes from Old French where *til* speaks of 'the lime trees'. Here we find Thorpe in Balne, a hamlet where the name has two quite separate origins. The element *thorp* is commonplace and speaks of 'the outlying farmstead'. Its frequency, it is one of the few suffixes often found

Kirk Sandall canal bridge, south of Barnby Dun – photograph courtesy of Wikipedia Creative Commons

on its own, is the very reason there is an addition describing 'the woodland clearing where beans are grown' from Old English *bean leah*.

Trumfleet comes from the same language, although *trun fleot* meaning 'the stream flowing in a circular course'. Clearly this makes no literal sense and thus should be understood differently. The usual meaning of *fleot* is 'tidal stream', the ebb and flow of the tide would explain the idea of flowing both ways, and yet this place is too far from the coast and the meaning is unknown. However we can speculate. Looking at the map there is a square formed by this and a second stream which unite to the southeast. Perhaps when the stream was named the two, which also run fairly closely to the northwest, were seen as flowing in two directions as both were known by the same name.

Barnsley

Recorded as Berneslai in Domesday, here a Saxon personal name is suffixed by Old English *leah* and speaks of 'the woodland clearing of a man called Beorn'.

Street names include Cheapside, a name found in many towns and cities which comes from *ceap* or 'market'. The word has come to mean 'cheap', and a bargain is something we would expect to find at our modern markets. The market is also mentioned in Market Street. Shambles Street, also still found in some towns and cities, most famously the example in York, comes from *sceamol* 'the bench for the display of goods' and a second reference to a market stall. This road was formerly known as Westgate as it led west out of the town and towards Manchester.

Town End speaks for itself as the outer limits of Barnsley. Pinfold Hill is a common street name, this speaking of where stray livestock were penned awaiting collection and payment of the fine by the owner. Back Lane is another common name describing the way into or out of the town. While clearly there is no 'front' or 'back' to a town, the reference is to the alternative or least used route. Old Mill Lane speaks for itself as the road to or passing where corn was ground. Star Line was named after the former Star Inn and is now a part of Regent Street. Similarly Cock Yard is a reminder of the Cock Inn, Wellington Street recalls the hero of the Battle of Waterloo, and Horns Lane is after the Horns Inn.

Pitt Street and Peel Square are named after former prime ministers William Pitt (either the elder or the younger) and Sir Robert Peel (most often said to be the founder of the modern police force). Queen Street was named in 1837, the year of the coronation of Queen Victoria. Earlier it had been known as Sough Brigg, from the River Sough, and shares this origin with Sowbridge Street. Church Street tells us it is home to the church of St John the Evangelist.

Not particularly 'new' today, and this is true of so many New Streets, it should be seen as meaning 'newer' and earlier known as Dawson's Lane after the family who lived here. Wager Lane is also from a surname, the last to live her was Mary Wager who died in February 1789 aged 72. In Shearshaw's Yard, previously known as Church Fold, shows this was land belonging to the church and several members of the Shearshaw family were buried here.

Rich Lane recalls Emor Rich, a former master at the Old Grammar School who died in April 1788. Adjoining the school was a house once occupied by a Mr Heelis who may have worked at the school, although there is no evidence to support this, he certainly gave his name to Heelis Street. Dog Lane is something of a mystery, perhaps known for hounds or maybe there was a dog-leg in earlier times.

Roper Street recalls rope makers Joseph and James Mallinson, these the last recorded although there may have been others here as there was in Billy West's Yard, he also documented as a rope maker. Armitage's yard recalls John Armitage, not born in the town but who came here and bought into several businesses. His hard work for Barnsley saw him made a town commissioner. Park Street remembers the estate known as Locke Park of the Locke family. Jumble Lane means what it says, this was where bits and pieces were sold and/or collected. Tattershall's Corner takes a famous horseracing name for this was where betting men were want to gather.

Old Town is self-explanatory, this from *ald tun*. Pog Moor features the early dialect word *pog* to speak of the 'bog' here. Wilthorpe comes from *wil thorp*, this the 'wild outlying farmstead' and showing what was once a satellite agricultural community has been abandoned. Beevor Hall was a building which probably never existed, the origin of *beofor hol* speaks of 'the hollow where beavers were seen'.

Public houses of Barnsley begin with the White Bear which, as with most featuring coloured animals, is heraldic and points to a connection to the earls of Kent. In the Manx Arms we see a clear statement of a connection to the Isle of Man, however the 'arms' in this case is used to denote a public house, the most common use of the phrase.

Obviously it is the product which is the most important item and one of the most ingenious names advertising same is the Tut 'N' Shive. These are two wooden pegs which are driven into the barrel to allow the contents to be drawn off. The most obvious is that used as a tap, yet it would not allow a smooth flow were it not for the second peg which always plugged the hole created at the top of the barrel when it lay on its side. Removing the top peg first allowed air to replace the liquid drawn off and was replaced afterwards.

Trees are large visual pointers to a location, hence they are used extensively in pub names. The same is true for the Oakleaf, the slight alteration distinguishing it from others. Location is behind the legal names of the Chambers and the Courthouse Station, although perhaps these may also fall under the heading of trades along with the Moulders Arms and the Commercial Inn, the latter showing this was where travelling salesmen would meet after hours of business in times when travel was much slower and communication not instant unless face to face.

Transport has provided the Keel Inn, these flat-bottomed boats were a common sight around the Yorkshire estuaries and coastline, while the Morning Star has been used for the names of a number of railway locomotives. The Corner Pin refers to the game of skittles and the most difficult pin to knock down. At the Dove Inn we find a religious reference, this a reminder of the first bird to return to the ark after the flood.

In the Prince of Wales Feathers is a visual symbol pointing to the heir to the throne, most often the future Edward VII whose term tenure saw more pubs built than at any other time mostly due to him holding the title longer than anyone before him. That record has been beaten by the present incumbent, ironically more pubs have closed during Prince Charles' time than any of his predecessors.

The Elephant and Castle is a name which continues to be explained as a corruption of the Infanta de Castile. Doubtless this creative etymology will continue despite this being far from the true origin. As with many pub names this is heraldic, representing a trade. Here we see the crest of the Cutler's Company, the link to elephants through the use of ivory as handles, while the 'castle' is always depicted on the back of the animal and is actually a howdah.

Barugh

From Old English *beorg* and meaning simply 'hill', this name is recorded as Berg in 1086, Bercam in 1090, Berig in 1243, as Berghe in 1297 and as Barghe in 1379.

Locally we find Gawber, this from Old English *galga beorg* or 'the hill of the gallows'.

Bawtry

A name from Old English *ball treow* tells us this was 'the tree rounded into the shape of a ball'. This name, recorded as Baltry in 1199, probably points to a tree cut into such an unnatural shape to act as a marker.

Flowing through here to the Trent, the River Idle is one of the few river names which can still be recognised as referring to 'the slow river'. This comes from Old English *idel*. Martin Hall Farm is from *maere tun* or 'the boundary farmstead'. Gally Hills is from *galga hyll* and a warning this was 'the hill where the gallows were erected'. The field names of Burrass Close and Far Burrass share an origin in Middle English *burgeis* and a reference to 'the citizen or burgess'.

Public houses include the Turnpike, a reminder of how tolls were once payable and such barriers were sensibly placed alongside roadside inns. The Granby was named to recall the soldier John Manners, Marquis of Granby. The loyalty shown by his men was rewarded when they left the armed forces when he set them up as inn-keepers.

Bawtry's history has seen several quite distinct periods. First the Romans were resident here, then the Saxon settlement was at the southernmost boundary of the kingdom of Northumbria. By the time the Vikings were in residence Bawtry was within the borders of neighbouring Nottinghamshire. The present site is not that of the original Bawtry, which has migrated as populations have changed. However the border with Nottinghamshire is still within a stone's throw of the villagers, which is shown in the first house on the Great North Road being named Number One Yorkshire, when travelling north.

Beauchief

Listed in the twelfth century as Beuchef, this is Old French *beau chef* and tells of 'the beautiful headland or hill spur'. Until the twentieth century Beauchief was a part of the traditional county of Derbyshire.

above: Beauchief Abbey is a common sight for those who play the golf course
right: Abbeydale Hall at Beauchief has been converted into exclusive living accommodation

Beighton

Early records of this name include Bectone at the beginning of the eleventh century. Here Old English *bece tun* tells of 'the farmstead by the stream'. The church of St Mary the Virgin has parts dating from the twelfth century.

Bessacarr

Listed as Beseacra in 1182, here is Old English *beos aecer* telling us it was 'the cultivated land where bent grass grows'.

St Wilfred's church, Bessacarr

Billingley

Listed in 1086 as Bilingeleia, here a Saxon personal name precedes Old English *inga leah* and describes 'the woodland clearing of the family or followers of a man called Bill or Billa'. The name has also been transferred to Billingley Green, from *grene* or 'village green', and Villingley Ings, where *eng* means 'meadow'.

Field names include Bond Branne Hill, brom hyll being 'the hill where broom grows'. Brecks Close is from *braece* or 'the land newly broken for agriculture'. Gorr Broad puts together *gara* 'a triangular plot' and *braedu* 'the broad strip of land'. Hunger Hill comes from *hungor hyll*, meaning exactly what it says and a warning this land was unproductive. Sailes comes from *salh* or 'willow'. Water Slades is derived from Old English *waeter slaed* meaning 'the wet valley'.

Birley

The Domesday record of Birle shows this to be Old English *burh leah* and refers to 'the woodland clearing by the stronghold'. There is a Frecheville Arms and a Frecheville estate developed in the 1930s. However the Frecheville family were lords of the manor of Staveley, there is no evidence they were ever directly influential in Birley.

Blaxton

Seen as Blacston in a document of 1213, this comes from Old English *blac stan* and describes 'the black boundary stone'.

Locally we find Blaxton Balk, where *balca* speaks of 'the boundary strip'. Blaxton Banks features the addition of *banke* or 'slope'. Blaxton Grange shows this was land set aside as 'an outlying monastic farm'. Here we also see Acomb, from *ac holmr* 'the dry land in marsh by the oak trees'. Ling Common features the element *lyng* or 'heather'.

Historically Blaxton was a part of the soke of Doncaster. To define a soke is difficult for it seems to have a different meaning in differing places. In some places it appears to be a judicial reference, in others the right to levy fines or forfeits, and even the etymology is no help as it comes from the word *secan* 'to seek'. What seems certain is it came to our shores with the Vikings.

Bolsterstone

A name which, when defined, shows this place was known for its gruesome history above everything else. From Old English *bolster stan*, this name refers to 'the stone on which criminals are beheaded' and recorded as Bolstyrston in 1398.

The Castle Inn is about the only reminder of the former Bolsterstone Castle, correctly a fortified manor house. Built by the Earl of Shrewsbury in the fourteenth century, stone from the building has been used to produce Castle Cottages and Porters Lodge, both dating from the nineteenth century.

Bolton upon Dearne

One of three Boltons in the traditional county of Yorkshire, each the 'farmstead with a special building' from Old English *bothl tun*. This example is recorded as Bodeltone in 1086, the addition refers to the River Dearne which is discussed under its own entry.

Here we find the name of Goldthorpe, nothing to do with precious metal but a name speaking of the *thorp* or 'outlying settlement of a man called Golda'.

Bradfield

The earliest surviving record of this name is as Bradesfeld in 1188. This is a fairly common combination of two Old English elements, *brad feld* speaking of 'the broad open land'.

Two small streams meet here, the Black Brook has a name describing the apparent colour of the river but not a reference to the water but to the bed of the stream being muddied rather than stony before its confluence with Ewden Beck. The latter is listed as Aquam de Udene in 1280, as Uden in 1420, with the modern Ewden for the first time in 1628. This is an example of back-formation from Ewden, itself from *iw denu* or 'the valley marked by yew trees'.

Load Brook is named twice, with *lad* 'watercourse' with *broc* 'brook'. This runs into the Loxley, a name from British or Celtic *loksko* and meaning 'crooked'. A former name of the Loxley is the Steyne, derived from *staene* meaning 'stony'. In turn these waters empty into the Rivelin, a name from *riveling* and meaning 'rivulet', a tributary of the River Don and a name seen only since 1310 and suggesting the name is not the original.

Bradfield can claim to be one of the largest parishes in England in terms of area. At 35,600 acres there can be few larger.

Braithwell

Listed as Bradwelle in 1086, this is from Old English *brad wella* or 'the broad stream'.

Minor names here begin with Austwood House, the building named after 'the east wood' which existed here well before the house was ever seen. Mickelbring takes its name from 'the great slope'; from *brom leah* or 'the woodland clearing where broom trees grow' comes Bramley; and 'the *by* or farmstead of a man called Helgis' is the origin of Hellaby.

A royal charter of 1289 granted the village a weekly market each Tuesday, with an annual fair lasting a remarkable eight days. These were rescinded long before the survey of 1652.

Bramwith, Kirk

This name is found as Branuuet in 1086, as Branwyth in 1200, and as Kyrkbramwith in 1341. Here the basic name is a combination of Old English *brom* and Old Scandinavian *vithr* telling of 'the wood overgrown with broom'.

Later we see the addition of *kirkja*, Old Scandinavian for 'church'.

Here we find Braithwaite, this from Old Scandinavian *breithr thveit* to speak of 'the broad woodland clearing'. Kirkhouse Green began as *kirkja hyrst grene* to tell of 'the village green at the wooded hill with a church'. In the case of Dowborn Bank and Palmer Lane we find two names derived from the surnames. There is no record of a Dowber

Top Lane Bridge, Kirk Bramwith – photograph courtesy of Wikipedia Creative Commons

Hellaby Hall, Bramley – photograph courtesy of Wikipedia Creative Commons

family here, nor any of Palmer, which may point to Middle English *palmer* meaning 'a pilgrim'.

Branton

This name is from Old English *brom tun* or 'the farmstead where broom grows'. The name is found in Domesday as Brantune.

Brierley

A name which describes 'the woodland clearing where briars grow' from Old English *brer leah* and recorded as Breselai in 1086 and as Brerelay in 1194.

Minor names include Grimethorpe, where Old English *thorp* follows a common Saxon personal name to speak of 'the outlying farmstead of a man

The mining wheel at Brierley – photograph courtesy of Wikipedia Creative Commons

called Grimr'. Ringstone Hill has seen little change since the Saxons referred to the *hring stan* or 'the ring of stones', just as the name suggests today. Folly Hall is not a folly in the modern sense, this comes from the family named Folyot who once lived here. Their surname means 'foolish' and has led to the use of 'folly' to mean a construction serving no purpose.

Public houses include the Robin Hood. If we were to believe this legendary outlaw visited every inn taking his name, the man would have spent so much time on the road he would never have had the chance to fight the Sheriff of Nottingham. Indeed the vast majority of these inns were built long after his lifetime, making his visits impossible. The name was adopted for the figure on the sign dressed in green, for the name of Robin Hood is far more romantic than the true reference to the woodsman or forester, although their role in village life was infinitely more important.

Similarly the Three Horse Shoes is another common name, an image used heraldically by the Worshipful Family of Farriers although that idea will have come from blacksmiths advertising their trade alongside the pub – it makes sense if the name is seen as a question, the solution lying alongside the public house.

Brinsworth

Recorded as Brinesford in Domesday, this name is from a Saxon personal name and Old English *ford* and tells of 'the ford of a man called Bryni'.

Held by the Percys, De Busli, Reresbys, Saviles, and Finches families, the Fullertons leased the coal seams beneath the village in the late nineteenth century. Their contribution to the area is marked by Fullerton Drive.

Brodsworth

A name from Old English *worth* following a Scandinavian or Saxon personal name and describing 'the enclosure of a man called Broddr or Brord'. This name appears in Domesday as Brodesworde. The basic name has also been transferred to Brodsworth Holt, *holt* meaning 'copse', and from *wudu* comes the self-explanatory Brodsworth Wood.

Brodsworth Hall – photograph courtesy of Wikipedia Creative Commons

Local names include Scausby, where Old Scandinavian *by* follows a personal name and tells of 'the farmstead of a man called Skulkr'. Conversely Pickburn is an Old English name, one referring to this as 'the pike stream' and a reference to good fishing, not for sport but for food. Chapel Plantation is self-explanatory, this managed woodland being near the place of worship. Clappers Hill features the Middle English *clapere* or 'rabbit hole'. Long Lands comes from *lang land* and describes 'the long agricultural land', ie: much longer than it is wide. Mellinder Dike is derived from a surname, itself from someone who was 'a milliner'. Norman Hill can also be traced to a personal name or surname. Roman Ridge is derived from the way otherwise known as Ermine Street or Great North Road. Windmills Drain began as *wind myln hyll* and is indeed an early example of 'the hill with a windmill'.

Field names begin with Beat Close, *bete* telling of the 'rough sods' of earth. Bowards Yard comes from *bur geard* 'the enclosure with a cottage'. Custs

Yard began as 'the enclosure of a family called Cust'. Foss Field was named from 'the ditch' at the side of the field to aid drainage and from *foss*. Havercroft started as 'the smallholding where oats are grown' from *hafri croft*.

Pease Close is from *pise clos* or 'the smallholding where peas are grown'. Spetchill Close comes from *spec hyll*, quite literally 'speech hill' and showing this was where regular meetings were held. Stocking Close features the element *stoccing* or 'tree stumps'. Tup Close links Old English *tup* with Old French *clos* to speak of 'the enclosure where the ram is kept'. Yaud Hill Field comes from *jalda hyll* or 'the hill where nags are kept'. Lastly we see Saint Foyn Leys, not a religious reference but from the '*leah* or woodland clearing where sainfoin grows'. This was a plant gathered for forage.

Burghwallis

Listed as simply Burg in 1086 we find the later form of Burgwaleys in 1283. The original name is from Old English *burh* or 'stronghold', later the addition is manorial from the Waleys family who were here by the twelfth century.

The local name of Robin Hood's Well remembers one Robert Hode, recorded on a deed dated 1322. While tradition maintains this name shows

above: Burghwallis sign; *right*: St Helen's, Burghwallis

the famous outlaw visited here, the deed is dated more than fifty years before the earliest mention of the man from Sherwood Forest.

Burghwallis Hall was the scene of a most macabre series of events in March 1934. Firstly a mother smothered her sleeping son and hid his body in a closet. When her husband discovered his young son's lifeless body he exacted his own retribution by murdering the murderess and then committing suicide. There have been unconfirmed reports of the sounds of a music box accompanying the sound of a young boy. Furthermore there is also mention of a man in grey and a woman of diminutive stature walking the corridors.

C

Cadeby

Here is an Old Scandinavian name featuring a personal name and *by*, speaking of this place recorded as Catebi in 1086 as meaning 'the farmstead of a man called Kati'.

The basic name has been transferred to a steep area known as Cadeby Cliff, and Cadeby Riding where *rydding* means 'clearing'. We also see Cadeby Rattles, the addition from Rattles Hill and speaking of 'the hill infested by rats'. Grey Stones Plantation is an area of managed woodland named from *graeg stan*, literally 'grey stones' and probably a marker. Spen Dike comes from *spenne dic*, a small channel across the Don at an angle and thought to refer to 'the dike which spans' and probably an aid to irrigation and/or drainage. Skitholme is a field name from *skitr holmr* and describing 'the dung covered dry land in a marsh'.

Campsall

Campsall sign

Documented as Cansale in 1086, Camasella in 1137, Camshale in 1142, Camsala in 1150, Campshale in 1158, Kamishalla in 1225, Camesale in 1264, as Campsale in 1294, and as Kempsall in 1547. Here Old English *halh* follows a British or Celtic *cambo* and gives 'the nook of land at the crooked place'. The first element is related to the Proto-Welsh *cambeis* and the Saxon personal name Cam.

Church of St Mary Magdalene, Campsall

Cantley

Listed as Canteleia in 1086, the Domesday record suggests this is 'the woodland clearing of a man called Canta' where the Saxon personal name is suffixed by Old English *leah*.

Street names come from a selection of themes. The usual trees are seen in Acacia Road, Birch Road, Lilac Grove, Maple Avenue, Sycamore Grove, and Willow Avenue, while those of Craven Close, Highbury Avenue, Ewood Drive, Villa Park Road, Hillsborough Road, Sincil Way, Ninian Grove, Anfield Road, Ayrsome Walk, and Goodison Boulevard, use the more unique football grounds.

From Old English *beos aecer* comes Bessacarr or 'the acre of land frequented by bees'. Branton began as *brom tun* or 'the farmstead where broom grows'. When it comes to pronunciation Brockhole has changed little since Saxon times when it was *brocc hol* or 'the hollow frequented by badgers'. Kilham Farm perpetuates the place name from *cyln ham* and telling of 'the homestead with a kiln'. When it comes to High Ellers and Low Ellers we find the origin to be Old Scandinavian, here *elri* speaks of 'the elders' with self-explanatory additions.

Carbrook

Seen as Kerebroc in the early thirteenth century, most likely this is Old Scandinavian *kjarr broc* and describes 'the stream in a marsh'. Alternatively the first element could represent the earlier Celtic river name.

Carlton

Listed as Carlentone in the Domesday record of 1086 and as Karletona in 1155. Here Old English *ceorl tun* speaks of 'the farmstead of the ordinary freemen'.

Locally we see Athersley Wood, this coming from a personal name and Old English *leah* to tell of 'the woodland clearing of a man called Aethered'. Boulder Bridge is a later name, this from Middle English *bulder brycg* describing 'the bridge built of boulders'. Wharncliffe Wood shows this was once a possession of the earls of Wharncliffe, while St Helen's Farm was once held by the Prior of Monk Bretton and a place of pilgrimage with several springs dedicated to St Helen. The name of the well and saint were once used by the local secondary school.

Cawthorne

Records of this name begin with Caltorne in 1086, the Domesday form shows this to be the '(place at) the cold or exposed thorn tree' from Old English *cald thorn*.

Barnby Hall occupies land originally known as 'the *by* or farmstead of a man called Bjarni'. Rons Cliff has evolved from *hraefn clif* and tells us it was once 'the bank where ravens were seen'. Womax has come from the local dialect describing 'the holm oaks'. Cannon Hall was named for being held by the monks of Monk Bretton. Jowett House and Rowling House were named for former residents, the Jewet or Jowet family and the Rawlin family respectively.

Delightful Cawthorne signpost

Three names are lasting reminders of the work carried out here. Cinder Hill is the easiest to see as where cinders and/or slag from a kiln was once deposited. Beet House gets its name from *bete*, Old English for 'sods' and telling us these larger clumps of wet soil were removed to produce a more easily managed agricultural land. In the case of Bark House Lane we see Middle English *bark hus* showing this was where a 'tan house' stood. Doubtless locals were more than aware this was where leather was tanned, for this process resulted in a veritable assault on the olfactory sense.

A tributary of the River Dearne, Daking Brook is unrecorded until the seventeenth century, although we should not discount the very real possibility the river was recorded by a name which we simply have not associated with this watercourse. This name is thought to come from the personal name Daykin with *broc* or 'brook'.

Cathorne's museum, built for a visit by Queen Victoria, may not be the biggest in the county but may well boast some of the most unusual exhibits. On show are a number of smoking aids used by native Americans, a local cow sadly born with two heads, and the boot of a man worn when he was struck by lightning.

Chapeltown

Here is a name seen as Le Chapel in the thirteenth century and as Chappeltown in 1707, this is from Middle English *chapel* meaning 'the place by the chapel'.

Clayton

A second place of this name in Yorkshire, this example recorded as Claitone in 1086. Here the Domesday record points to Old English *claeg tun* and 'the farmstead on clayey soil'.

The local name of Frickley speaks of 'the *leah* or woodland clearing of a man called Fricca'.

Conisbrough

Another name featuring the Old Scandinavian *konungr* this time with the Old English *burh* and telling of 'the king's fortification'. This name is recorded as Cunugesburh in the early eleventh century and as Ciningesburg in Domesday.

The local name of Butterbusk is clearly stating this region was known for its dairy farming, indeed the name speaks of 'the bush on rich pastureland'. In Firsby we see an Old Scandinavian *by* telling of 'the farmstead of the Frisians'. These people heralded from the coastal regions of what is now Germany and the Netherlands.

At its peak there were more pubs known as the Red Lion in England than any other. Numbers once exceeded six hundred but in the twenty-first century this has fallen. For such a common name, it has only been seen since the seventeenth century and, rather ironically, the most common pub name in England is representative of the monarchy of Scotland.

It began as a device on the coat of arms of the most powerful man in England in the fourteenth century, John of Gaunt. His vast riches would, in today's money, be worth in excess of £100 billion, making him one of the twenty wealthiest people in history. His successors went on to become kings of England through lines resulting in Henry IV and Henry VII and, of course, their descendants. While few pub names can be traced to the fourteenth century and directly to John of Gaunt himself, his red lion symbol did come to England through a more circuitous route. Of his fifteen named children, four with Katherine Swynford were later legitimised by royal decree, taking the surname Beaufort. Their descendants thereafter became every ruler in Scotland from 1437 and, following the accession of the House of Stuart in 1603, every monarch of England.

Cudworth

Seen as Cutheworthe in the twelfth century, this represents a Saxon personal name and Old English *worth* and tells of 'the enclosure of a man called Cutha'.

Minor names include Pen Hill Lane, with Old English *pening halh* referring to 'the nook of land worth one penny'. The same language has also given us Sid Copp, *set copp* telling of 'the seat-shaped hill'; *smael brycg* refers to 'the narrow bridge' rather than the modern name of Small Bridge; while Belle Green is from *belle grene* is understood as 'land given over to raise money for the maintenance of the church bells'.

Pub names include the Pinfold, which takes an old minor place name speaking of the place where stray livestock was brought awaiting collection and payment of the fine by the owner.

D

Dalton

Another example of a name also found in North Yorkshire, one found in Domesday as Daltone. As with its namesakes this represents Old English *dael tun* and means 'the farmstead in the valley'.

Foljambe Drive was named after the family who purchased this estate in the fourteenth century.

Darfield

This is a name found as Dereuueld in 1086, the Domesday listing pointing to an origin from Old English *deor feld* and telling of 'the open land frequented by deer or other animals'.

Here the minor place names begin with Edderthorpe, here the suffix of *thorpe* follows a personal name to tell of 'the outlying farmstead of a man called Eadric'. Both Great Houghton and Little Houghton share an origin of 'the farmstead at the nook of land'. Here both places come from Old English *halh tun* and with obvious additions. Aldham has not changed since Saxon times, for this is Old English ald ham or 'the old homestead'.

In Wombwell we see the common suffix *wella* following an unusual personal name for an English place name as it comes from Old High German and speaks of 'the spring of a man called Wamba'. Hemingfield adds *inga feld* to the personal name to speak of 'the open land of the family or followers of a man called Hymel'. Note this was named for his family or followers, not the man himself, and thus was probably named posthumously in Hymel's memory. Smithley is easy to see as 'the *leah* or woodland clearing of the smith or metalworker'. Tyers Hill is a later name, one derived from the surname

Darfield village club – photograph courtesy of Wikipedia Creative Commons

Tierce which is not seen until the Middle Ages when the language had evolved to Middle English.

Locals enjoy a drink at the Cross Keys, one of a number of public houses named after the symbol of St Peter. If religion and public houses seem an odd mix, we should realise the church was the second largest landholder in the country, after the monarch.

Darton

Listed as Dertun in the Domesday survey of 1086, this represents Old English *deor tun* and describes the 'farmstead for deer' (a deer park).

Mapplewell has changed little since Saxon times, *maple wella* describing 'the spring by the maple tree'. Pye Wood retains the early name of the magpie, until comparatively recently simply known as a *pie*. Staincross comes from *stein cros*, this 'stone cross' showing this was once of great importance in this region as this marked the meeting place for the wapentake - an administrative district of the Danelaw and the equivalent of the Saxon hundred.

At the Rose and Crown we find the epitome of a patriotic pub name, where the former represents the nation and the latter its monarchy.

Dearne (River)

A river name recorded as Dyrne in 1154, as Derna again in 1154, and as Derne in 1340. These forms are rather late and it seems highly improbable this name was not already at least a few centuries old by the twelfth century. Hence it is not clear if this is an early Saxon name or of late British or Celtic derivation. This is important for it changes the meaning drastically. Should this be from Old English *derne*, then this would speak of the 'hidden, secluded one', yet the earlier British *duro* meaning 'fort' would refer to a fortification standing on its banks.

Note this latter origin in particular would only apply to a very short stretch of the river. This is true of all rivers to some degree, for the young vibrant fast-flowing torrent soon slows and meanders across the plain before slowing to a crawl and influenced by the tide in old age. At every stage there are names describing the river and, as travel far from home was the exception, it stands to reason there must have been several names for each water course depending upon where the river was flowing. That very few rivers have more than one name is, as with the Derwent below, purely by chance when the cartographer asked the name of the river he had just drawn on the map.

The water quality has fluctuated over the years. By the late nineteenth century, despite attempts to rectify the problem, the sewage from the homes on the banks and effluent from forty-four mills had polluted the river to such an extent it was declared dead. The situation had not improved by the 1960s and despite the many measures being taken to prevent pollution had scarcely improved by 1974. In the 1980s trout were released in the upper reaches and were a self-sustaining population ten years later but even today the lower Dearne still cannot be considered clean.

Denby

Found as Denebi in 1086, as Denbye in 1304, and as Dembye in 1562, here is a name featuring the common Old Scandinavian suffix *by* to speak of 'the farmstead of the Danes'. The name was later transferred to the satellite settlements of Lower Denby and Upper Denby.

Local names include Bagden, where early forms do not make it easy to distinguish between meanings of 'the bag-like valley' or 'the valley of the bdgers'. Deffer Hill is a transferred name, the hill overlooking 'the deep ford'. Strines features the element *strind* or 'stream'. Trister Hill comes from Middle English *trystor* and speaks of 'the hunting station'. While Dunkirk is not seen until the map of 1822, undoubtedly this remoteness name had been around for some time.

Denby was the birthplace of John Flamsteed, the first Astronomer Royal. Opposite St Mary's church is his memorial garden, where the night sky can be seen mapped out at all times of the day and night.

Derwent (River)

This is one of four rivers of this name in England, all of which have identical meanings. Found as Deruuentiosis fluvii in a document from the eighth century, this listing speaks of 'the river where oak trees grow'. This is a perfect example of how many modern river names are named purely by chance. It is impossible to believe oak trees grew along every inch of the banks of this river and on both sides, thus we must assume this applied to certain areas along the banks and there was every likelihood a different characteristic gave a different name elsewhere along its course. Hence this was named Derwent purely by chance, reliant on an early cartographer asking the question of the name of the river at a certain point along its course.

Dinnington

Listed as Dunintone in Domesday, this is derived from a Saxon personal name and Old English *ing tun* and describes 'the farmstead associated with a man called Dunna'.

Lorden's Hotel, Dinnington St Johns – photograph courtesy of Wikipedia Creative Commons

Howard Avenue recalls former lords of the manor the Howard family, earls of Norfolk. Athorpe Lodge, Athorpe Lodge and Athorpe Road share an origin in the family who purchased the estate in 1678, although then the family name was still appearing as Athropp, John Athorpe the first lords of the manor to live here. The local Falcon Inn derives its name from the bird found in the Athorpe family's crest. St Leger Way and St Leger Road are named after Anthony St Leger, major landholders in Dinnington by the eighteenth century.

Dodworth

Seen as Dodesuuorde in Domesday in 1086, this comes from a Saxon personal name and Old English *worth* and refers to 'the enclosure of a man called Dodd or Dodda'.

Minor names include Hugset Wood, where the Middle English element *side* follows a personal name to give 'the hillside of a family called Hugger'.

Champany Hill is another later name, this Old French *champaigne* referring to 'level open country' and also the origin of the place name which gave a name to the famous French drink. Around the same time the Savile family were living at the region still known as Saville Hall. Ratten Row is much earlier, this from Old English *raten raw* 'the row of rat-infested houses'. This must have been true of most houses at the time, hence the name was probably given by those who disliked or looked down on the inhabitants.

At the Travellers Rest the message is one of welcome, an invitation to any passer-by to enter. In the Horse and Jockey we see a reminder there

top right: Dodworth sign;
above: Cusworth Hall – photograph courtesy of Wikipedia Creative Commons

were once many more horseracing courses than today. This made sense as before the coming of the railways horses would walk to the course, thus limiting the distance it could travel and still be expected to perform. Sporting links can also be found in the Pheasant, this Asiatic bird was brought to Britain to stock the country estates but soon escaped and is now happily living wild. Its startling plumage made it a good subject for a pub sign and probably indicated a link to shooting and may have indicated it was on the menu.

Don (River)

Although the earliest records are all post-Conquest, this name is certainly pre-Saxon and possibly pre-Roman. Found as Don in 1135, Douni in 1161, Done in 1171, Doon in 1275, Dun in 1300, Done in 1350, Dunne in 1463, and Dune in 1485. All suggest a British word *dana*, itself from a root *dan* meaning simply 'water, moisture, river'. There is also a Little Don River, which simply shows it is a tributary of the Don, and an Old River Don, little more than a drain today but was once the course of the Don itself.

Three tributaries share a common origin. The River Went is fed by both the Went Beck and the Little Went, all from a British or Celtic *eto* with Welsh *gwen* and together understood as 'the pleasant one', a reference to its mainly calm and predictable flow.

Doncaster

Listings of this name include as Doneceastre in 1002 and as Donecastre in 1086. This name refers to 'the Roman stronghold on the River Don', where Old English *ceaster* follows a river name discussed under its own entry.

Streets of the town begin with Duke Street, formerly known as Roper's Walk as it was where rope was produced and its name changed to honour the Duke of Wellington. Known as Scote Lane in 1454, Scot Lane comes from *scot* a term meaning 'tax, toll or penalty' and still used in the phrase "to get off scot free". The old town gates are represented by Sunny Bar, originally known as Sun Bar.

Fishergate was recorded as Vico Piscatorum in the Middle Ages, referring to 'the way of the fisherman', with later references to High Fishergate and Low

Fishergate and a single listing as Friendly Street. Laithgate describes 'the way with or near a barn' from Old Scandinavian *hlatha* and Old English *barn*. Frenchgate shares this suffix, this being easily seen as 'the street associated with Frenchmen' thought to be Norman craftsmen. It was once recorded as *Vicus Franciscus*, but was not named for the Franciscan Friars as they are documented as arriving in 1307 but this name is first documented in 1270. This is not the case with Greyfriars Road, this indeed from the colour of the habits worn by these monks.

Baxtergate is a corruption of Bakster and listed as Vicus Pistorum during the Middle Ages, this being 'the street of bakers'. During the reign of Elizabeth I trading of geese took place in what is now known as Goosehill. Bower Fold remembers the Bower family, first mentioned in the parish register in 1565 when John Bower married Alice Margrave in December of that year. Ten Pound Walk leads to the engine sheds of the GNR, the name referring to the rent payable to Doncaster Corporation to provide access to the sheds across their land.

Silver Street was only known as such from the eighteenth century, prior to that is was known as Back Lane (among others). While Silver Street is the new name, Golden Street is the old name. Both renamed around 1792 by Alderman B. Earnshaw and one R. Dey and held to be "Proof these two gentlemen delighted in the coin of the realm". Golden Street became Factory Lane, named as this was where the Reverend E. Cartwright established his wool and cotton manufactory in the early nineteenth century. The same man is remembered in Cartwright Street, honoured as inventor of the power loom.

Other lost street names include Magdalens, clearly a reference to the church dedicated to St Mary Magdalen. Meal Lane was a reminder of where meals was sold, the Shambles named for the meat and fish market, the insides of the meat and fish littering the streets leading to the modern use of 'shambles'. Ratten Rawe shows this was 'the rat-infested row of houses'. Scestang Lane is derived from *sa stong* meaning 'the vat and pole', understood to show where buckets were hung from a pole and carried from their landing place at the river. West Laith Gate speaks of itself as 'the road of the western barn'. Station Road is much later and cannot have been named before the railways arrived.

Clock Corner comes from the clock hung on the building at the corner of High Street and Baxter Gate around 1837. Hall Gates is a reminder of one of the town's former gates. St James Street remembers the chapel of St James, the first record of which dates from 1689 when it was noted as being a ruin. Bennetthorpe was the location of the houses where one Joseph Bennet lived from 1778. Cleveland Street remembers William Henry Vane, who followed his father as the 3rd Earl of Darlington in 1792, and was created Marquis of Cleveland in 1827.

With St Sepulchregate the reference should be to St Pulcheria herself, or indirectly through a Carmelite order of nuns dedicated to this saint. However there is not any mention of either in existing documents. One theory, which does seem to be stretching the facts a little, involved the Barons Revolt against King John in 1215, a date famous for the signing of *Magna Carta*. Some have suggested such an order of nuns helped fund the raising of the defences in Doncaster against the king and his forces. These suggestions do not clarify whether these funds were given willingly or not.

Minor names here begin with Balby, an Old Scandinavian personal name with *by* telling of 'the village of a man called Balli'. Hexthorpe is from the same tongue and gives 'the outlying farmstead of a man called Hegg'. We also find Old English as the origin of Loversall and Wheatley, giving 'the *halh* or nook of land of a man called Leofhere' and 'the woodland clearing where wheat is grown' from *hwaete leah*, respectively.

The River Cheswold connects the Don with a former course of the Don near Crimpsall. The meaning is not altogether clear but it is certainly an example of back-formation. Here the first element is certainly *cese* and is followed by either *wald* to give 'the woodland near where cheese is produced' or, if the second element is *wella*, speaks of 'the cheese vat or well'. The latter would be more likely to refer to a cheese-like shape of the vat rather than an area known for its cheeses.

Pubs include the St Leger Tavern, recalling the most famous race run at Doncaster and the oldest of the five English Classics. Unlike the other races, which are open to horses of either colts or fillies, this is open to both. Not only is it the oldest but is also the longest and always the last on the calendar each year. The Winning Post is similarly a reference to the sport of kings, as does the Paddock and also the Beechers Brook, despite it being the best-known fence on the course at Aintree. The Flying Childers was the name of an eighteenth century horse trained by and named after Sir Hugh Childers.

The Woolpack recalls the wool trade, the basis of our nation's wealth and subsequent empire, and how the pubs of England acted as delivery and collection points before the coming of the canals and railways. Taken from the coat of arms of the Weavers' Company, the Leopard is another reminder of the importance of cloth. To some degree the Nags Head is related, not as a trade but when travellers could hire these small sturdy animals to transport them to the next in a chain and thereby make their way across the country. Transport has also given us the more obvious name of the Railway, another in the Station, and again in the Loco, one of the most famous locomotives in history in the Mallard, and also in the Great Northern.

One of the most famous stages in world history is found in two pubs. The Pilgrim Fathers names those who sailed to the New World, while the Mayflower remembers the vessel in which they travelled. The Millstone shows the importance of grinding flour to feed the people, less obviously the Staff of Life also refers to bread. The Golden Fleece is symbolic of the Knights of the Golden Fleece, who were tasked with protecting the church.

The Two Palfreys were horses of smaller stature, making them ideal for women. Two such animals were given to King John in the eleventh century, in return he granted Doncaster a market. The John Bull Inn is a patriotic image created by John Arbuthnot in the early eighteenth century.

Dore

Found as early as the late ninth century exactly as it appears today, this name comes from Old English *dor* and describes the '(place at) the narrow pass'.

The *Anglo-Saxon Chronicle* records Dore as being the site of the victory by King Egbert of Wessex over King Eanred of Northumbria and effectively uniting Saxon England.

E

Ecclesall

Listed as Eccleshale in 1200, as Eccheleshalla in 1218, and as Ecclesall Byerlow in 1587, here British or Celtic *ecles* precedes Old English *halh* to refer to 'the nook of land with a church'. The sixteenth century record includes the addition of Old Scandinavian *byjar log* telling us this is 'a township'.

Locally we find a tributary of the Rivelin, the Black Brook named for the colour of the river bed rather than the water. The Meers Brook, a tributary of the Sheaf, derives its name from *mere* or 'pool'. Also emptying into the Sheaf, the Limb Brook may come from *hlynn* 'torrent' or if this is an example of back-formation then *lin* or 'flax' will be more plausible. Porter Brook is not recorded until 1587 and must be an example of back-formation and shares an origin with Porterfield in an (unrecorded) surname.

Bell Hagg can be taced to *bel haga* to show this was 'the enclosure marked by a beacon'. A name from *burh leah* is seen on modern maps as Birley Meadows, this being 'the woodland clearing of the borough'. Brincliffe unites Old English *brende clif* and describes this as 'the steep place cleared by burning'. Broom Hall was built at the 'place where broom grows'. Fullwood comes from *ful wudu* 'the wood at the dirty or muddy place'.

While Sharrow Head certainly comes from *scearu* and means 'the share of common land', this term is also used to refer to 'a boundary' and 'a district' but, with no second element, it is difficult to know just what is being described. However it clearly refers to an important place and thus perhaps all three are relevant in referring to a meeting place.

Greystones is from *greg stan* 'the grey stones' most often referring to a boundary marker. A synonym has given us Whirlow, where *har hlaw* speaks of 'the mound at or near the boundary marker'. Hallam uses *halh* in a plural form to tell of 'the nooks or corners of land'. Heeley is an unusual modern form of the common combination *heah leah* referring to 'the high woodland clearing'. Hillsborough had been 'the hills' for a long time until the fairly modern addition of 'borough'. Millhouses comes from *myln hus* or 'the houses at or near the mill'.

Owlerton has nothing to do with birds, this is from *alor tun* or 'the farmstead at the alder trees'. In Upperthorpe we see 'the *thorpe* or outlying farmstead of a man called Hopere', where the personal name is derived from an ancestor's trade as 'a cooper'. Walkley began as the *leah* or 'woodland clearing of a man called Walca'. Whiteley Wood began as *hwit leah wudu* or 'the wood of the white or bright woodland clearing'. Yarncliffe is derived from *earn clif* or 'the steep place frequented by eagles'.

Little Sheffield is not exactly from Old English but has been transferred here from the name of the modern city, perhaps suggested by the first residents having come here from the original Sheffield settlement. Old Scandinavian has given us Crookes, from *krokr* or 'the nook or corner of land'. With Crookesmoor the name tells us of 'the moor near Crookes'. Snaithing Farm and Snaithing Wood are derived from *sneithing* or 'a detached piece of land'.

Ecclesfield

Here is a name telling of 'the open land near a Romano-British Christian church'. This name comes from Celtic *egles* and Old English *feld* and is recorded as Eclesfeld in the Domesday record of 1086.

The Blackburn Brook, a tributary of the River Don, is derived from the colour of the stream bed, which is muddy, not stony.

Edlington

Listed as Ellingtone in Domesday and as Edelington in the late tenth century, here is 'the farmstead associated with a man called Edla', where the Saxon personal name precedes Old English *ing tun*.

Minor names begin with Blow Hall, this built at a place known for its 'conical pile of unhewn stones'. Broom House stands at 'the house where broom grows'. Cockhill House comes from *cocc hyll* or 'the hillock on the hill'. Martin

Ecclesfield church – photograph courtesy of Wikipedia Creative Commons

Wells Farm adds *wella* or 'stream' to the personal name Martin. Similarly Peter Wood Close uses *wudu* or 'woodland' with an obvious personal name. Stubbins Hill comes from *stubbing*, literally 'a clearing' and probably saying it is cleared of tree stumps, much as the term is used today.

Field names begin with Bully Tree Close, where Middle English *bolas* points to 'the wild plum'. More fruit is seen in Folks Orchard is a lasting reminder of a family named Foukes who once worked this area. Piper's Acre also features a surname, here Middle English *pipere* means 'piper' and lead to the name Piper.

Pub names include the Royal, a clear patriotic name, and the much more interesting White Greyhound. This animal was the constant companion of the First Viscount Molesworth, who erected a monument to the animal when it died. When walking the gardens of the state one day the dog stopped his master from entering the privy by pulling on his coat tails. The master asked a gardener to investigate and the poor man was shot dead by a man hiding there, probably waiting to rob the house.

F

Fenwick

Seen as Fenwic in 1166, this comes from Old English *fenn wic* and describes 'the specialised (probably dairy) farm in a fen or marsh'.

Fenwick London Hill has no connection with the nation's capital. Here the addition to the village name has three elements, *land hyrne hyll* describing the 'cultivated land in the bend at the hill'. Lawn comes from *launde* meaning 'forest glade', while Lady Thorpe is from *hlaefdige thorp* or 'the outlying farmstead at the tumulus'.

Finningley

A name from Old English *fenn inga leah* speaking of 'the woodland clearing of those associated with the fenland'. This name appears as Feniglei in 1086 and as Feningelay in 1175.

Local names include Auckley, where the listings make it impossible to be certain of the personal name preceding Old English *leah*. Hence this could be anything from 'the woodland clearing of a man called Alca' to 'the woodland clearing of a man called Alha'. In Blaxton we see Old English *blaec stan* where the 'black stone' would have been an old boundary marker.

Holy Trinity church, Finningley – photograph courtesy of Wikipedia Creative Commons

At the Harvey Arms we see a pub named after the family who came here in the seventeenth century as lords of this manor.

Firbeck

Domesday lists this name as Fritebec in the twelfth century, a name from Old English *fyrhth* and Old Scandinavian *bekkr* showing this is 'the stream from the wooded hill'.

An aerial view shows one field has a noticeably oval shape, this was once the private racecourse of a local racehorse owner. Anthony St Leger was the man who started the running of the race named after him in 1776. One of the five English Classics of the flat racing season, it is always run at Doncaster and yet locals fervently believe it had been run for some years prior to this at Firbeck.

Fishlake

Listed as Foscelac in Domesday, this is 'the fish stream' from Old English *fisc lacu.*

Here we also find minor names, including another named from a stream. Sykehouse comes from *sic hus* and tells of 'the house by the stream'. Assenthorpe Green features the suffix *thorp* and describes 'the outlying farmstead of a man called Aestan'. From Old English *wurm leah* or 'the woodland clearing frequented by snakes' comes a name which has been transferred to Wormley Hill.

Public houses show links to the land and to the church. In the Hare and Hounds we are reminded of the blood sport where greyhounds were tested in their abilities when chasing the hare, the focus being on the chase rather than capture. Not being particularly close to the sea, the Old Anchor represents the church, where Christianity is said to be the anchor through life's storms.

Perhaps a future local name is suggested by the Cockatrice of Church Street. This is from a local story which speaks of the churchyard being the haunt of this fabulous beast. However those who should hear it are warned they are doomed never to sleep again, although we are not told just how we should recognise the sound of the cockatrice were we unfortunate enough to witness it.

G

Gawber

With the earliest record as Galgbergh in 1304, this comes from Old English *galga beorg* and describes 'the hill of the gallows'.

Gildingwells

Seen as Gildanwell in the thirteenth century, this is probably a name from Old English *gyldande wella* describing 'the gushing spring'.

Gleadless

Recorded as Gledeleys in the thirteenth century, this name comes from Old English *gleoda leah* and describes 'the woodland clearing frequented by kites'.

Goldthorpe

Found in the Domesday record as Goldetorp, here is 'the outlying farmstead of a man called Golda', where the Saxon personal name is suffixed by Old Scandinavian *torp*.

Greasbrough

Again the earliest record available comes from Domesday, appearing as Gersebroc. This name comes from Old English *gaers broc* and refers to 'the grassy brook'.

Gunthwaite

Found as Gunultwayt in the twelfth century, with records of Gunnildethwayt in 1257, Gunyldthwayte in 1284, as Gunnilthuayt in 1318, and as Gomuldthwayth in 1549, this name refers to 'the woodland clearing of a woman called Gunhild'.

Burnt Cote is still very much self-explanatory, this being the site of the 'burned cottages'. Ellison Spring, which features a family name, uses Old Scandinavian *spring* which means 'plantation'.

H

Hallam

Found in 1086 as Hallun, the Domesday record is not enough for us to choose between the two possible explanations. Either this is the '(place at) the nooks of land' from Old English *halum*, the plural form of *halh*, or, should this be Old Scandinavian *hallr*, the '(place at) the rocks'.

Hampole

From Old English *hana pol* and meaning 'the pool frequented by cocks (of wild birds)' this appears as Hanepole in the Domesday record of 1086. With the difference in languages, the Germanic tongue of the Saxon tenants and French speaking Normans, the great tome is a notoriously unreliable source of proper names and thus the name could be referring to 'the pool of a man called Hana'.

As with many places in this general area, Hampole is keen to show a connection with the legendary outlaw of Sherwood Forest. A nearby stream is known as Robin Hood's Brook, a name undoubtedly influenced by Robin Hood's Well, a local water source which is marked by a stone covering designed by John Vanburgh. However there is no water to be drawn here, for the 'well' has been moved to its present location as it was standing in the path of the dual carriageway of the A1 road. Hampole also boasts a Little John's Well, undoubtedly named to complement the other.

Handsworth

A name from a Saxon personal name and Old English *worth* points to 'the enclosure of a man called Hand', this is found as Handesuuord in the Domesday survey of 1086.

The common name from *brom leah* or 'the woodland clearing where broom grows', has now been taken for Bramley Hall. Gleadless is less obvious, for early forms do not show whether this is from *glaed leah* and 'the bright woodland clearing' or *gleoda leah* 'the woodland clearing frequented by kites', this referring to the birds of prey.

Handsworth's traditional sword dance is still performed. Eight men perform the dance which ends with their long steel swords being interlocked so as to be held aloft by just one of their number. It is claimed the dance is of ancient origins, although written records date from the nineteenth century and the music seems to date from a similar era.

Harthill

Documented as Hertil in Domesday, a name derived from Old English *heorot hyll* which tells of 'the hill frequented by harts or stags'.

Hickleton

A name found as Icheltone in 1086, Hikalton in 1164, as Hikiltun in 1285, Ykelton in 1291, Hekilton in 1304, and as Hikleton in 1409. There is no doubt this comes from Old English *hicel tun* to describe 'the farmstead frequented by woodpeckers', however it should be noted the first element is also used as a personal name.

Hooton Levitt

Listed as simply Hotone in the Domesday record of 1086, the basic name is derived from Old English *hoh tun* and describes 'the farmstead on a spur of land'. The addition is manorial, a reminder of the Livet family who were here by the thirteenth century, which distinguishes it from the following entries.

Hooton Pagnell

A second 'farmstead at the spur of land' from Old English *hoh tun*, this name appears as Hotun in 1086 and as Hotton Painel in 1192. Again the addition is manorial, this time the Painel family who were here by the eleventh century.

Here we find the hamlet of Stotfold, a name describing 'the enclosure known for its stud farm'.

Hooton Roberts

Another Old English *hoh tun* or 'farmstead at the spur of land', this example has records of Hotun in 1086 and as Hoton Robert in 1285. This later listing is a reminder this place was a possession of a man called Robert in the thirteenth century.

Houghton, Great

Found as Haltun in 1086 and as Magna Halghton in 1303, this is a common name and comes from Old English *hoh tun* or 'the farmstead at a hill spur'. The addition is Latin *magna* which, as the modern form tells us, means 'great' and distinguishes it from the following entry.

Houghton, Little

A second *hoh tun* or 'the farmstead at the hill spur'. Here the name appears as Haltone in 1086 and as Parva Halghton in 1303. As with the previous entry the addition is Latin, here *parva* is seen today as 'little'.

High Hoyland sign

Hoyland, High

The basic name comes from Old English *hoh land* speaking of 'the cultivated land of or by a hill spur' with the addition *heah* or 'high'. This name is recorded as Holand in 1086 and as Heyholand in 1283, with the addition distinguishing it from the following entries.

Hoyland Nether

A second 'cultivated land of or by a hill spur' from Old English *hoh land*. Here the additional *neotherra* means 'lower' and describes it in relation to the previous entry. Records include Hoiland in 1086 and as Nether Holand in 1390.

Hoyland Swaine

The final 'cultivated land of or by a hill spur' from Old English *hoh land*, here the addition is manorial and is from the twelfth century landholder called

Hoyland Swaine church

Swein or Sveinn. The name is recorded as Holande in 1086 and as Holandeswayn in 1266.

Early records show High Lee has evolved from *haelig leah* 'the holy woodland clearing', probably an indication some religious ceremony was held here but no records of same have survived. Cat Hill is a reminder of the residence of the Catling family. Similarly Renald Lane remembers a local family recorded as both Rennel and Reynold. Mustard Hill means exactly what it says, this place was associated with the condiment and derived from Middle English *mustard.* Pickness Hall took the name of the location, it describing 'the pointed headland'. Scout Dam took the name of a local feature, itself from Old Scandinavian *skuti* meaning 'projecting or overhanging cliff or rock'.

Hunshelf

This name is derived from Old English and means 'the *scelf* or steep land of a man called Hun' and is listed as Hunescelf in 1086, as Hunschelfe in the thirteenth century, as Hundeschelff in 1365, and as Howenchelfe in 1440.

Snowden Hill speaks of itself as 'the bare or bald hill', one lacking vegetation. Dyson Cote is easily seen as being named for the Dyson family. However while we know the origin of Unsliven Bridge is as it was recorded previously as Unshriven Bridge, we have no idea as to the meaning of this word and thus the meaning of the name, and of the word, remains a mystery.

I

Ingbirchworth

Records of this name include Berceuuorde in 1086 and as Yngebyrcheworth in 1424. The original name is Old English *birce worth* or 'the enclosure where birch trees grow'. Later we see the addition of Old Scandinavian *eng* or 'meadow'.

The local name of Annat Royd uses the Middle English *rod* or 'clearing' after a female personal name which began as Annot. Maze Brook forms the boundary between here and neighbouring Thurlstone, this from *maere broc* and describing 'the boundary stream'.

Scout Dyke is a very modern name, one coined by locals when this was used as a base by US troops during the Second World War. This thereafter gave its name to the reservoir, although Scout Dyke Reservoir can sometimes become so low the old Summer Ford Bridge is revealed once more. The irony being this bridge replaced, and was named after, the ford which was only passable during the summer when water levels were low enough.

K

Kexborough

Domesday records this name as Cezeburg, with a later twelfth century record of Kesceburg. Here is a name from an Old Scandinavian personal name and Old English *burg* referring to 'the stronghold of a man called Keptr'.

Birthwaite Hall took the existing place name when it was built, this from Old Scandinavian *birki thveit* or 'the woodland clearing where birch trees grow'. Low and Upper Swithin feature the element *swithinn* which tells us this was 'land cleared by burning'. Scrat Haigh Wood unites Old Scandinavian *skratti* and Old English *haga* and tells us it was known as 'the enclosure haunted by goblins'.

Snapethorpe is an odd name which merits some discussion. As mentioned elsewhere in these pages a *thorp* is an 'outlying farmstead', one beginning as a satellite of the earlier community and undoubtedly first used only in the warmer months, ie: during the growing season. Founded to feed a growing population it would have grown at a similar rate and eventually have become a permanent home to some. However here the first element contradicts this for Old English *snap* describes 'poor pasture'. Hence it seems the land would not have produced good crops either and raises the question as to why it was settled. It is true that derogatory terms were used in place names as insults to those who worked there but this makes no sense here.

Kilnhurst

Seen in the twelfth century as Kilnhirst and derived from Old English *cyln hyrst*, this is 'the wooded hill with a kiln'.

Kiveton Park

Recorded as Ciuetone in the Domesday record of 1086, here is 'the farmstead by a tub shaped feature' from Old English *cyf tun*.

L

Langsett

Seen as Langeside in the twelfth century, this is from Old English *lang side* and describes 'the long hill slope'.

Belle Clive is the modern form of 'the *clife* or bank of a man called Billa'. Coming from Old English *hord aern*, what is recorded as Hordron today speaks of 'the store house'. In Uskers we have a name from *hus kjarr* speaking of 'the house at or by a marsh'. Swinden began as *swin denu*, meaning 'the valley where hogs are seen'. Laund is derived from Old Scandinavian *launde* speaking of 'the forest glade'. Damstakes is easy to understand when we learn this was where 'stakes were used to dam a river'. Finally Alderman's Head was land of the *aldormann*, while this was the origin of the title it did not mean the same thing but once described 'the chief officer of the shire'.

Laughton en le Morthen

Domesday lists this name as Lestone in 1086, a name from Old English *leac tun* and speaking of 'the farmstead where leeks or garlic is grown', a common name which should be seen as more than just those two vegetables for undoubtedly others would have been grown here and was thus the beginning of market gardening. Later we find the addition, no surprise for such a common basic name, found as Latton in Morthing in a document dated 1230. This refers to it being 'in the district known as Morthen' and from Old Scandinavian *mor thing* 'the assembly point on the moor'.

Some sources cite the addition as being from Old French for 'the place of death' and held to be a reference to the Battle of Brunanburh. However the Normans did not arrive until some 150 years after the battle, thus this seems most unlikely. Others point to this as where the last major battle between the Saxons and the Vikings and the Celtic peoples to retain (or regain) control of England and Wales, and where some 50,000 individuals are said to have lost their lives. The Celts or Britons certainly fought no such battle as late as the tenth century, and losses of 50,000 lives would seem more than a little excessive.

Local names include Thurcroft, which has changed little since its beginnings as 'the croft of a man called Thori'. Slade Hooton is another of the places derived from Old English *hoh tun* or 'the farmstead at the spur of land', here with the distinguishing addition of *slaed* or valley'. Letwell is a late name, one from Middle English *lette* and describing 'the obstruction', one found here in the *wella* or 'stream'. Langold is derived from Old English *land halh* or 'the long nook of land', this giving some clue as to position of those who described the place such for, from a compass point ninety degrees either way, it would have appeared (and thus named) as 'broad'. Throapham has two elements, a plural *thorp* with the later addition of *ham*, giving 'the homestead at the outlying farmstead'.

Letwell

Found as Lettewelle in the middle of the twelfth century, this name comes from a combination of Middle English *lette* and Old English *wella* referring to 'the spring or stream with an obstructed flow'.

Loversall

Recorded as Louershale in 1086, the Domesday record shows this to be from a Saxon personal name and Old English *halh* and refer to 'the nook of land of a man called Leofhere'.

M

Maltby

From Old Scandinavian, there are two possible origins for this name. Either this is where the personal name precedes *by* and 'the village of a man called Malti' or comes from *malt by* and 'the farmstead where malt is made', ie: an

Roche Abbey, Maltby – photograph courtesy of Wikipedia Creative Commons

early brewery. This name appears as Maltebi in the Domesday record of 1086.

Some thirty years ago the traditional game of beck ball was revived. Any rules seem arbitrary at best, but could be described as akin to rugby but played out between the banks of the local stream or beck.

Marr

Little change for this name since the Domesday record of Marra, and no change whatsoever from the original Old Scandinavian *marr*. This name refers to the '(place at) the marsh'.

Melton, High

Found as Middeltun in 1086, the Domesday records shows this was 'the middle farmstead' from Old English *middel tun*. The addition comes much later, a self-explanatory element expected for such a common place name.

Mexborough

Seen as Mechesburg in 1086, the Domesday record shows this to be 'the stronghold of a man called Meoc or Mjukr', where the Old English *burh* follows a Saxon or Scandinavian personal name.

In Denaby Rage we see the first name from *denn by* or 'the farmstead by a woodland pasture', together with the fairly uncommon *raec* or 'reach' used in the sense of 'headland'.

Public houses include the Star Inn, showing a link to the church through its most famous story and the Star of Bethlehem. The church is also behind the Ring O' Bells Inn, perhaps inviting the bell ringers to imbibe there after practice. The Sportsman Inn is another invitation, a general message to those who play sports both indoors and outdoors. The Boy and Barrel is a cryptic advertisement for the product, this symbolising Bacchus, Roman god of wine.

Monk Bretton

Seen in 1086 as Brettone and in 1225 as Munkebretton, this is from Old English Brettas *tun* and 'the farmstead of the Britons' with the addition of *munuc* showing an association with the monks of Bretton Priory.

Lund Wood is something of an oxymoron, for the second element is superfluous as Old Scandinavian *lundr* already describes a 'copse' or 'small wood'.

The local is the Full House. This name is clearly inspired by the hand in poker – where the five cards include three of one value and two of another – but the real message is a packed establishment, and a busy pub means a popular pub and hopes to attract business by the name alone.

Mosborough

Seen as Morsesburh in the early tenth century and as Moresbury in Domesday, here is 'the fortified place in the moor' from Old English *mor burh.*

Battles have inspired pub names, most notably Waterloo and Trafalgar. However the Battle of Alma, the first major engagement of the Crimean War, has also spawned a surprising number named the Alma and was also used as a street name. Furthermore the soldiers who fought here brought the name back and it became a Christian name for girls overnight. The river name comes from the Tatar word for 'apple', although whether the parents knew this is debatable.

Moss

Whether this comes from Old English *mos* or Old Scandinavian *mosi* this name describes 'the swamp or bog'. The name is first seen in 1416 as Mose.

Minor names here include Moss London Hill. Not a reference to the largest city but an addition from *land hyrne* or 'the cultivated corner of land'. Moseley Grange shares its first element with the village name. Here *mos leah* points to 'the swamp at the woodland clearing'. In Flashley Carr we see the same Old English suffix, for here *flask leah kjarr* speaks of 'the marsh at the pool in the woodland clearing'.

N

Norton

One of the most common place names in the country, and surprisingly one which has not attracted a second element to distinguish it from others. This is also true of early forms, this name seen as Nortone in 1086 and as Nortuna in 1135, and coming from Old English *nord tun* and speaking of 'the northern farmstead'.

Local names include Stygate Lane, this from Old English *stig gata* which describes 'the gate or gap at the path'. Spittlerush Lane is thought to to come from *spittle risc* to tell us this was 'the rushy place where spittle grows'. While spittle, also seen as saliva, was certainly a plant despite the common use of these terms to describe a digestive fluid. However we have no idea just what this old plant name refers to, the vegetation offering no clue.

O

Oughtibridge

Seen in 1161 as Uhiuabrigga, here a Saxon personal name and Old English *brycg* tells us of 'the bridge of a woman called Uhtgifu'.

Oughtibridge sign

Owston

Found in Domesday as Austhun, this is a combination of Old Scandinavian *austr* and Old English *tun* and refers to 'the eastern farmstead'.

Minor place names start with Carcroft, a common Old Scandinavian name speaking of 'the smallholding at the marsh'. Blacker Green is a name given to the village green, it from *blaec kjarr* or 'the black marsh'. Felhurst Bridge is derived from *fjol hyrst* and tells of 'the wooded hill near the plank bridge'. Rockley is from *hroc leah* or 'the woodland clearing frequented by rooks'. From *sur pece* comes Sourpiece Wood, the modern name defining the place name perfectly.

Shirley is an Old English place name from *scir leah* or 'the bright woodland clearing'. Cockshaw Dike is from *cocc sceaga* or 'the copse where woodcock are found'. Morley Well features a surname, while a Danish personal name has given us Turnholme from 'the *holmr* or dry ground in a marsh of a man called Tumbi'. Stream Dike, a tributary of the River Don, comes from *stream dic* a self-explanatory name where the first element is surprisingly rare.

Oxspring

Domesday's record of Ospring in 1086 and as Oxspring for the first time in 1154 point to a meaning of 'the spring frequented by oxen' from Old English *oxa spring*.

Local names include Roughbirchworth, this from *ruh birce worth* speaking of 'the rough enclosure near the birch trees'. Bower Hill is derived from *bur* meaning 'dwelling', a name which tells us very little compared to Walk Mill Bank, where *walc myln* refers to 'the fulling mill' and where cloth was processed. Wraith House is one of the earliest records of the modern use of *wraith* to describe 'a spectre' and pointing to a haunted house.

Oxspring sign

P

Penistone

Domesday lists this as Pengestone in 1086 with the later record of Peningeston in 1199. Here the original Celtic *penn* (meaning simply 'hill') was misunderstood by the Saxons who added Old English *ing tun* to refer to 'the farmstead by the hill known as Penning'.

Cubley puts together the Old English elements *cub leah* to speak of 'the woodland clearing of the cattle stall or shed'. Stotter Cliff also speaks of livestock, here *stod fald* was 'the enclosure for horses'.

Pilley

Recorded as Pillei in 1086, this is from Old English *pil leah* describing 'the woodland clearing where stakes are obtained'. Rather than seeing this as where the stakes were cut, remember this would have been in use for some time before the name would stick and the nearby woodland would soon be consumed, this should be seen as where they were stored ready to be sold or bartered for other goods.

R

Ravenfield

Listed as Rauenesfeld in Domesday, this is an Old English place name with two possible meanings. Either this is hraefn feld 'the open land frequented by ravens' or 'the open land of a man called Hraefn', if the first element is used for a Saxon personal name.

Rawmarsh

The earliest surviving records of this name are as Rodemesc in Domesday and as Rowmareis at the end of the twelfth century. Here is 'the red marsh' from Old Scandinavian *rauthr* and Old English *mersc.*

In the Fighting Cocks we see a reminder of the popular sport of cock fighting, one which attracted crowds from far and near and saw vast amounts of money change hands in bets. The Earl Grey is not an advert for tea, but for the former prime minister.

In the local church of St Mary's we find a place of worship completed in 1838, although there had been a church here since Norman times. The present building has seen several stages of rebuilding. As recently as 1977 paving in the sanctuary was replaced and the opportunity taken to honour the name of Canon Scovell, a man who also has a street named after him.

Riding (East, North and West)

The old divisions of the traditional county of Yorkshire, no longer used in any official capacity since the administrative changes which created North Yorkshire, South Yorkshire, and West Yorkshire. The name refers to the county being split in three parts for it comes from Old Scandinavian *thrithjungr* and telling of 'a third part'.

Rossington

Seen as Rosington towards the end of the twelfth century, this is a combination of Celtic *ros* with Old English *ing tun* and describing 'the farmstead associated with the moor'.

In the minor name of Littleworth we find a name which has clearly been changed deliberately and comparatively recently, although it is uncertain exactly when or by whom. Early records show this name came from an Old English word *scyttels* and akin to Old Norwegian *skutil* or Old Swedish *skyttel*, all referring to a pole or bolt used to fasten a panel and block (or open) a gap – fundamentally an early form of a gate as we would know it today. These early forms show the original place name would have been Shittleworth, which reveals why the name will have been changed.

The local named the Styrrup is simply a different spelling of the place a rider puts his feet when on horseback. A different way of describing the long relationship between the horse, its rider and the public house.

Rother, River

This tributary of the Don is found as Roder in 1164 and as the modern form as early as 1539. Despite the lack of records prior to the twelfth century there is no doubt this comes from the British tongue spoken by the Celtic peoples. Here *ro dubro* speaks of either 'the chief river' or 'the full river' depending upon how 'high' is being used.

Its twenty-seven miles were devoid of life for much of the twentieth century, indeed this was known to be the most polluted of all the tributaries of the River Don. Coking plants, manufacturers of chemicals, and terribly inefficient sewage plants all discharged into these waters. By 1994 plant and invertebrate life had recovered to a point where restocking of fish could begin and within two years some of the populations were already self-sustaining.

Rotherham

Found as Rodreham in 1086, the Domesday record speaks of 'the homestead on the River Rother'. This comes from a combination of a Celtic river name, discussed under its own entry, and Old English *ham*.

Street names include Church Street, clearly named for the parish church, but which was known as Ratten Row telling it was it had long been known as 'rat infested'. The rats will have been attracted by the easy pickings from the adjacent square once known as the Shambles. This area was where butchers sold meat at market stalls and will have slaughtered the animals close by so as to make transporting the carcasses so much easier.

Back to religion for Domine Lane, this the Latin for 'the Lord' and taken from the Domine Well which once erupted here. College Lane was named from the medieval College of Jesus, which occupied the northern side of College Lane between the street and the River Don. College Street was formerly known as Jesus Gate, both clearly from the College of Jesus. Howard Street takes the name of the dukes of Norfolk, the family had ironworks in the town by the seventeenth century when it was producing 450 tons of iron per annum. Effingham Street recalls the earls of Effingham, on whose lands were leased a slitting mill and a rolling mill.

Local names dating from the Middle Ages, and much earlier, can still be seen around the town although their definitions would be difficult to recognise in these early descriptions of the locality. For example there are the topographical features described by names such as Canklow, from Old English *canc hlaw* or 'the steep rounded hill or mound'. Dalton is 'the farmstead in the valley or dale' from *dael tun*. Greasbrough uses *burh* in the later sense to talk of 'the grassy manor', with Masbrough using the earlier *burh* in telling of 'the fortification on the boundary'.

Not only natural features but man-made sights have contributed to minor place names. Such as Scholes, a reference to a place known for its 'sheds'. Similarly what began as 'the barns near the wood' is now known as Woodlaithes. However Orgreave is rather different for this was 'the pit from which ore was dug'. In Catcliffe the sighting of feral cats led to this being known as 'wild cat bank' and nothing to do with the wild animal found north of the border.

People have left their name in the landscape, too. However it is unlikely any were ever aware they had been immortalised in the landscape. Here we see 'the ford of a man called Bryni' represented by Brinsworth. Badsley was known as the *leah* or 'woodland clearing of a man called Baeddi'. While Tinsley gets its name from the *hlaw* or 'mound or a man called Tynni'. Yet it is not only men who are remembered for at Kimberworth we see the *worth* or 'enclosure of a woman called Cyneburg'. In truth there are probably more personal names given as male which are female, yet differences in masculine and feminine forms are difficult to see today, much as the Saxons would find it hard to differentiate between our names Michael and Michaela, Michelle and Mitchell.

Family names are much easier to see, although these are always much later as surnames were not used until the arrival of the Normans. In Rotherham we find Herringthorpe, this being 'the outlying farmstead of the Hering family', who were recorded here for the first time in the twelfth century. The Barbot family, first recorded in a document dated 1316, were landholders at what is now known as Barbot Hall.

Pub names begin with the Alma Tavern, named after the first battle of the Crimean War fought on September 20th, 1854. A combined Anglo-French force outnumbered the Russians by almost two to one and were victorious on a day when some 9,000 lost their lives. The battle was named after the river and was quickly used as a pub name and a street name, with some soldiers returning home to christen their daughters Alma.

The Sitwell Arms remembers the family who owned the Renishawe estate when the pub was first opened two centuries ago. Other heraldic images are found in the Cutlers Arms, nearby Sheffield is famous for its cutlery manufacturers. A modern location reference is found in the Millmoor, inviting supporters for this is also the name of the local football club's ground. The White Lion is probably a reference to the dukes of Norfolk, although it is a common element in the arms of many families. No doubting the origin of the Shakespeare Arms.

In the Pewter Pot we see an advertisement for the product through the receptacle from which it is quaffed. Of course this is traditional coal mining country and the Davy Lamp honours the means of lighting underground when working which greatly reduced the dangers of gas explosions. Transport has given us the Mailcoach, from when it was all transported across land by coach and horses, while the coming of the railways has provided the Railway, the Sidings, and the Midland Inn. The Sir Colin Campbell recalls the field-marshal whose military career is best remembered for his role in the relief of the siege at Lucknow, India in 1857.

The Comedian remembers a local man, Sandy Powell being a star of music hall and radio whose name is probably not quite as well remembered as his catchphrase, "Can you hear me, Mother?" The Three Tuns is not a direct reference to the product but heraldically linked to the Brewers Company and the Worshipful Company of Vintners. In the Lumley Arms we see a reminder of the family who owned this land and Durham Castle, the Earls of Scarborough. Don John Inn remembers the winner of the St Leger, the last classic of the season, at Doncaster in 1838.

Royston

Telling us of 'the farmstead of a man called Hror or Roarr', here Old English *tun* is preceded by a Saxon or Scandinavian personal name and is recorded as Rorestone in the Domesday survey of 1086. The Cross Inn took an existing name, one showing this where a cross was used as a marker.

Royston's history is nearly wholly agricultural, although for a time industry came in the form of a colliery, a clay works, brick works, and a coke works, although only the latter is still operating. These industries relied on first the canal and then the railway as transport links, both of which are also closed today.

Royston's church of St John – photograph courtesy of Wikipedia Creative Commons

S

Sandall, Kirk

Listed as Sandala in 1086 and as Kirke Sandale in 1261, here the original name came from Old English *sand halh* telling of 'the sandy nook of land'. Later we find the addition of Old Scandinavian *kirkja*, telling us this place had a church.

Shafton

Seen as Sceptun in Domesday and as Scaftona in the middle of the twelfth century, this represents Old English *sceaft tun* and describes 'the farmstead marked by poles'.

When places were named they were pointers to that place and thus, at least in the local area, fairly unique. Lidgate is derived from *hlid geat* and tells of 'the swing gate' and while such are commonplace today, when such was named we can be sure this was a very new development. Prior to this the gate as we know it was simply a removal panel placed across a gap. Quilby Lane comes from a combination of Middle English *qwy* itself from Old Scandinavian *kviga* and with the common suffix *by* speaking of 'the farmstead where heifers are reared'. The local pub is the Great Dane, the likely origin of which is a favourite dog.

Sheaf, River

A tributary of the River Don which joins it at Sheffield and gives the city its name. Listed as Scheve in 1183, Shaua in 1312, and as Sheaffe in 1700, this comes from Old English *sceath* meaning 'boundary'. This has been an important river over the years, forming the boundary between the Saxon kingdoms of Mercia and Northumbria and, in more recent centuries, the traditional boundary for the counties of Yorkshire and neighbouring Derbyshire.

Sheffield

Recorded as Scafeld in the Domesday record of 1086, here is 'the open land by the River Sheaf', the suffix being Old English *feld*. The river name, from Old English *sceath* meaning 'boundary', is discussed in full under its own entry.

Street names in the Steel City, as with any reasonably-sized settlement, represent a mix of the history of the place. The oldest are early place names or perhaps destination names, however these have little connection with what we see today. Likely the most interesting and informative streets take the names of the people who were living or associated with the area.

Councils like to retain existing names wherever they can. Often these are in the form of landholders or, in the case of Pitchford Lane, their place of work of Pitchford Farm. Alderson Road was named after the family who owned land around the Highfield area. Much earlier was a Saxon landowner by the name of Aldene who was known in the Ecclesfield area and who is seen in Aldene Road. In Barnforth Street we see the family who lived at High House, Owlerton from the early seventeenth century.

Bernard Road recalls Bernard Edward, the 12th Duke of Norfolk, who held this land. Birks Avenue can be traced to John Birks, the first of a family first seen here in a document dated 1657. Bright Street remembers the family who owned Carbrook Hall by the seventeenth century. Samuel Swann Brittain gave his name to Brittain Street. The Brownell family lived around what is now Brownell Street for several generations from the seventeen century. Burgoyne Road recalls the lords of the manor of Owlerton from the eighteenth century.

Cary Road takes the name of Henry Cary, Lord Hunsdon and lord of the manor of Eckington. John Challoner gave the land to the town in 1795 and was suitably rewarded by the name of Challoner Way. Cooks Road took the name of landowner Henry Cook. Creswick Street remembers Thomas Creswick, lord of the manor of Owlerton in the sixteenth century. Falconer

Lane recalls the family who are recorded here in a document dated 1553 but whose name was recorded with the different spelling of Fawkener.

Farmer John Fisher was well known in the nineteenth century, hence the naming of Fisher Lane. Similarly William Galley, another farmer, gave his name to Galley Drive. Haden Street is cut on land held by William Haden in the late nineteenth century. Harwood Street takes the name of one Mrs Elizabeth Harwood who owned the land, as did George Hutchinson hold what is now Hutchinson Road, and Elizabeth Inkersall held Inkersall Drive in the eighteenth century. Bower Road was named after the family of Samuel Saville Bower.

Lovetot Street remembers the family who held the manor of Sheffield for most of the twelfth century. Newton Lane took the name of Jonathan Newton, who sold this land in the nineteenth century. Shortridge Street remembers landowner John Shortridge, Terry Street is after farmer Stephen Terry, an old Sheffield family were chosen when naming Trippet Lane, and Trevor Young held land on what is now Young Street. However no landowners earned their name more than the Rawson family who were here for more than five hundred years before the naming of Rawson Street.

Perhaps the most common choices are the local councillors who, having previously served this area, are honoured by their successors. Auckland Avenue was named after Austin Auckland, who served as clerk to parish council for nineteen years from 1950. Collinson Road recalls Thomas Collinson, a man who served the community in several offices for more than forty years. Crowther Place is named after nineteenth century alderman William Crowther.

Developers have invested money and quite rightly have been honoured by the naming of streets in their honour. William Ainsley is remembered by Ainsley Road, Hammond Street is after Thomas Hammond, and Goulder Place remembers a family who included five men who virtually built Attercliffe during the expansion of the nineteenth century. Builders have been marked by the naming of such as Carson Road, after joiner and builder William Carson. Mason Crescent marks the work of the film of J. C. Mason Ltd. With Popple Street remembers nineteenth century builder and contractor Abraham Popple.

Of course the name of Sheffield is still synonymous with steel and there are a number of street names recalling the employers and their factories. Ashberry Gardens remembers Phillip Ashberry, a former town councillor who also founded the Britannia Metal Works. Engraver James Bagshaw lived and worked for more than forty years in what is now Bagshaw Road. Bingham Road was named after James Bingham, a firm of steel and file manufacturers. The delightfully-named Blonk Street is after Benjamin Blonk, founder of Blonk & Co, scissorsmiths.

John Rose Bradley, horn, haft and scale pressers, is seen in Bradley Street. George Cadman established the Cadman Street Works in 1751 and saw Cadman Street named to commemorate same. Cowlishaw Road recalls cutler Y. T. Cowlishaw. Cyclops Works remembers the Cyclops Works, founded here in 1845. Dixon Road comes from the cutlery firm of James Dixon & Son, while another cutlery works owned by the Ellin family from the eighteenth century gave a name to Ellin Street. Thomas Ellin founded the Vulcan Works, forgers and rollers seen in what is now Forge Lane. Mountford Croft was inspired by George Mountford, maker of scythes, and knives for hay and straw in the nineteenth century.

Ellison Street recalls file and steel manufacturer Joseph Ellison. Jowitt Road was named after the factory of George Jowitt, producers of grinding wheels. Nowill Court took the name of Thomas Nowill, well-known for his Britannia manufactory of metal spoons. Outram Road takes the name of file manufacturer Henry Outram. Pagoda Street is from the Pagoda Works, where umbrella ribs were produced. Blacksmith John Reney is seen in the name of Reney Road. Warris Place recalls Councillor George Warris, silver plate manufacturer.

Other local businesses and businessmen are seen in Aizlewood Road, named after the chairman of the Sheffield Brick Company, who owned the land, John Aizlewood. William Storm, a shareholder of the Sheffield Brick Company, is remembered by Storm Street. Barlow Street recalls brickmaker and contractor John Barlow. Former grocer and beer retailer Andrew Barretta inspired the name of Barretta Street. Bates Street remembers surveyor Henry Brian Bates. Chemical Yard was the site of the old Totley Chemical Works.

Reuben Clarke, a builder and contractor, gave a name to Clarke Square. A nineteenth century grocery business is remembered by Deakins Walk. Estate agent William Downing is remembered by Downing Lane. Pawnbroker John Eaton left his mark in Eaton Place. Fawcett Street was the location of a boarding and day school for young ladies in the nineteenth century run by Marianne and Martha Fawcett.

Grammar Street was once home to the Old Sheffield Grammar School. Hayes Drive recalls John Ibbotson Hayes, master at Mosborough school in the nineteenth century. Hides Street recalls Henry Timm Hides, former pawnbroker. Keyworth Road comes from Charles Keyworth, owner of the New Patent Plastic Brick Company. Linen draper and merchant Matthew Lambert gave his name to Lambert Street.

Lister Road is after Henry Pickering Lister, Sheffield accountant and estate agent. Merchant and landowner Samuel Mitchell is recalled by Mitchell Street. Builder and joiner John Henry Preston gave his name to Preston Street. Vernon Terrace took the name of William Henry Vernon, former grocer and beer retailer. Willey Street remembers William Willey, not only a landholder but also owner of the Falstaff Inn. Merchants George and Sarah Woodhead are marked by Woodhead Road. While grocer Thomas Wragg gave his name to Wragg Road.

More locals are seen in names such as Applegarth Close after Robert Applegarth, a pioneer trade union man who was secretary of the carpenters and joiners union. Bedford Street was named because the land was owned by the Royal Infirmary Board and remembers John Bedford who served on the board for more than fifty years. Blake Street recalls typefounder and Justice of the Peace Thomas Blake. Charles Ashmore Road was named after the artist who moved here from his native Birmingham.

Gibbons Drive recalls the wood carver Grinling Gibbons. Nineteenth century benefactress Miss Eliza Harrison is remembered by Harrison Road. Similarly Hollis Croft is after Thomas Hollis, whose charity work saw him create charities, a school, and almshouses.

Lieutenant-Colonel John H Leslie gave his name to Leslie Road, he being a solider in the Royal Artillery and a military historian. Another solider, Henry de Marcham, gave his name to Marcham Drive. His name lived on following his role as bowman at the Battle of Crecy in 1346. Much later came Sir Henry Clinton, soldier and officer at the Battle of Waterloo, after whom Clinton Place was named. Clerk to the Sheffield Board of Guardians for some forty-five years was enough to earn Joseph Spencer the naming of Spencer Road in his honour. Spurr Street was named after former resident Peter Spurr, while Campo Lane simply tells us football was played here, and Esperanto Place recalls 1974 and the venue of the International Esperanto Conference.

Religious links saw the naming of Dunkerly Road, the Reverend Daniel Dunkerley was at Loxley Chapel in the nineteenth century. Fulmere Crescent comes from William Fulmere, vicar of Ecclesfield in the fourteenth century. Odom Court remembers Canon William Odom, vicar of St Simon's in the late nineteenth century and who wrote extensively about Sheffield. Dr Thomas Sutton, vicar in Sheffield for some forty-five years, gave his name to Sutton Street.

Of course there are those which refuse to be categorised. Albanus Croft can be traced to Roman times, for he was a Roman foot soldier, born a member of the Sunuci tribe, whose name was found on a copper diploma uncovered by a ploughman at Stannington in 1761. Clearly the man had gained some reputation as the relic states no less than the Emperor Hadrian had granted Albanus Roman citizenship. Any pride he felt in the reward he received in his lifetime would surely be surpassed by being made virtually immortal when naming the street.

Aldine Court is an area associated with printing. In the late fifteenth and early sixteenth centuries Aldus Manutius had a printing press in Venice and is credited with the invention of italic type. Burnaby Street takes the name of soldier and adventurer Frederick Gustavus Burnaby who, in 1882, crossed the English Channel in a hot air balloon. It seems he did so as he had not been chosen to participate in the Egyptian campaign. Rollet Close recalls one Peter Rollett, the confidential secretary to Mary, Queen of Scots, whose only connection to Sheffield was his imprisonment here.

However surely the title of most bizarre reason for choice of street name belongs to Muskoka Avenue. The name was that of the home of a Chippawa Indian chief in Ontario, Canada and named Nesqua Ukee – a name which means 'not easily turned in battle'. It was chosen by developers Albert and Alfred Burnett after they visited the place and simply liked the name.

Local politics have given us Benson Road, after Alderman Joseph Benson who served as mayor in 1925; Beverleys Road comes from Councillor George Beverley; councillor and JP William Cockayne is seen in Cockayne Place;

Gainsford Road remembers Alderman Thomas Gainsford; Ironside Close was named after Alderman Isaac Ironside; and Maxfield Avenue is after former councillor Joshua Maxfield.

Mundella Place is after local MP Anthony Mundella; Neill Road recalls councillor James Neill; parish councillor John Newsham is recalled by Newsham Road; Nuttall Place takes the name of Alderman William Nuttall; Alderman George Saunders is commemorated by Saunders Place; former councillor James Staton left his mark in Staton Avenue; Roebuck Road remembers nineteenth century MP John Roebuck; while Jeffcock Road remembers William Jeffcock, Sheffield's first mayor.

National political figures were the inspiration for Aberdeen Lane, after George Hamilton Gordon, Earl of Aberdeen and who served as prime minister from 1852 to 1855. Acton Place is after Lord Acton, former Liberal MP. Asquith Road remembers another prime minister, Herbert Asquith held office from 1908 to 1916. A third gave us Balfour Road, Lord Balfour coming into the office in 1902. Royalty is always a favourite and has produced Adelaide Street and Queen Adelaide, consort of William IV.

Other national and international figures have given us Allende Way, honours Dr Salvadore Allende, assassinated former president of Chile. Bazley Road links the city to the Howard family, dukes of Norfolk, through the marriage of Ruth Howard to Gardner Bazley. Corby Street also connects to the dukes of Norfolk, another line of which lived at Corby Castle in Cumberland. Egerton Lane links politician Lord Francis Egerton with the dukes of Norfolk through the marriage of his daughter. Marcus Smith worked for the dukes from 1820 until his death in 1871 and gave his name to Marcus Drive.

Cambridge Street was named for the Duke of Cambridge, who laid the foundation stone to mark those who gave their lives in the Crimean War. Dugdale Drive remembers seventeenth century antiquarian Sir William Dugdale. Fairfax Road recalls General Fairfax, who commanded Parliamentary troops during the English Civil War. Exchange Gateway is after a point on the coaching routes where fresh horses were hitched up.

Landowners have left their mark on the map, too. Aldam Close remembers William Aldam Milner; Bradford Street comes from George Bradford; Brightmore Drive is from the family of William Brightmore and his daughter

Elizabeth; Burton Street took the name of the family who were lords of the manor from the early eighteenth century; John Candow is remembered by Candow Street; Edward Cheney gave his name to Cheney Row.

At first glance Conalan Avenue has no connection with landowner Thomas Osbourne, yet the two are connected. Mr Osbourne had two children but there was only one road to name, hence the names of daughter Constance and son Alan were combined. Cutts Terrace was built on what was previously the estate of George Cutts, who had his own metal manufacturing business. On the subject of Sheffield's industrial past, it's most famous manufacturing of cultery produced Cutlers Walk.

Darwin Close is after the family who held land here from the seventeenth century. Dyson Place remembers Miss Eliza Dyson, here in the nineteenth century. Fentonville Street was cut on what had been Fentonville Gardens, this the land owned by the Fenton family. The Dixon family held the land where Fielding Road was cut, the connection between the two names from the marriage of Florence Dixon to Sir Charles Fielding. Similarly Fitzmaurice Road recalls the marriage of the landowners, the Huntsman family, in the shape of Benjamin Huntsman to Anna Fitzmaurice in 1850.

Fox Lane takes the name of farmer Thomas Fox. Matthew Furness, owner of the Robin Hood Inn and the surrounding land, is recalled by Furness Close. Glave Street recalls the property of decorator William Glave. Charles Goodwin owned shops and the property at what is now Goodwin Road. Joshua Gray, well known as a manufacturer of scissors, is marked by Gray Street. Greaves Close is named after several generations of farmer names Greaves.

Land owned by the Watson family could easily have become known as the nepotism estate as all the roads were named after members of the family and include Hannah Road. Hawke Street recalls landowner Peter Hawke. Hawley Street took the name of the Hawley family. Heppenstall Lane was named after the delightfully-named Price Heppenstall. Hunstone Avenue recalls farmer Ernest Hunstone. Another farmer, Joseph Hunter, left his mark in Hunters Lane and a third in Jardine Close and Robert Jardine. Joshua Road took the name of Joshua Spooner. Tillotson Close is named after nineteenth century landowner John Tillotson. Tyzack Road takes the name of the family of Joshua Tyzack, a family of farmers.

Latham Square is after Thomas Latham. Lennox Road was cut on land owned by the Dixon family, in particular Lennox Dixon. Malin Road was named from Malin Bridge and ultimately from Malin Stacie, seventeenth century landowner. Newbould Lane took the name of William Newbould. Lead merchant John Nodder gave his name to Nodder Road. Nineteenth century solicitor Broomhead Colton-Fox married the daughter of Captain Staveley Sherde, leading to the naming of Sherde Road.

Builder and mason Henry Stothard is remembered by Stothard Road. Algernon Sydney, son of the Earl of Leicester, was a parliamentarian soldier at the Battle of Marston Moor and is remembered by Sydney Road. Joiner and builder Charles Taplin is recalled by Taplin Road. Tudor Place recalls one half of a partnership in a silver-plating business, Henry Tudor. Joiner and builder George Travis is recalled by Travis Place. Brick manufacturer J. Webster helped develop Webster Street. Whiting Street recalls a Mr Whiting, one of the creditors of the Meersbrook Tannery Company who petitioned for the company's winding up and whose name shortly afterward appeared on signs at the ends of this road.

Manufacturers off spades and shovels C. T. Skelton left their mark in the name of Skelton Lane. Builder T. Smith is remembered by Smith Street. Solly Brothers, a well-known family of iron masters, is marked by Solly Street. Tool manufacturers John Sorby and Sons left their name in Sorby Street. Stancer Street remembers grocer Joseph Stancer. Ironfounder Thomas Steade is remembered by Steade Road.

Revill Lane can be traced back to 1657 with the first record of freeholder John Revill. Nineteenth century clothier Benjamin Richards gave his name to Richards Court. Builder and contractor Harry Ripley is remembered by Ripley Street. Builder and joinjer David Rushby was responsible for some of the properties in Rushby Street. Former miner, later chairman of the miners' union, William Sewell gave a name to Sewell Road.

Penthorpe Close remembers early nineteenth century Chartist Thomas Penthorpe. Perkyn Road took the name of former resident John Perkyn. Polishing paste manufacturer Joseph Pickering is recalled by Pickering Road. Powell Street remembers James Powell who first developed. Raisen Hall Place took the name of Raisen Hall, named from a family recorded as Raysin, Rayson and Reason.

Nicholson Place remembers Benjamin Nicholson of Shoreham Steel Works. Former lord of the manor John Ormond gave his family's name to Ormond Close. File manufacturer Thomas Osborne built Osborne Cottage at where we now see Osborne Road. Parkers Lane takes the name of post office keeper Elisha Parker. Parkin Street was named after John Parkin of metal working company Blake and Parkin. Trustee John Pedley gave his name to Pedley Avenue.

Marples Close recalls John Marples, a prominent figure at the Sheffield Brick Company. Marsden Lane is a misnomer and was originally Masdin Lane after John Masdin, owner of a local firm producing pen-knives, pocket-knives and table knives. Knives were also produced by James Matthews, a man remembered by Matthew Street. Moncrieffe Road is after Miss Annie Moncrieffe Rundle, who lived here up to the early twentieth century.

Lister Avenue recalls G. R. Lister and Son, asphalters and road makers. Livesey Street recalls the Rev John Livesey, vicar of St Philip's Church for thirty-nine years. Similarly the Rev Alexander Mackenzie, vicar of St Paul's Church, gave his name to Mackenzie Street and the Rev Edward Mansel is recalled by Mansel Avenue. Logan Road remembers John Logan, railway contractor. One of the most laudable to be honoured is Henry de Marcham in Marcham Drive, an English bowman at the Battle of Crecy in 1346.

Ketton Avenue is linked to Frederick Walker, assistant managing director of the Ketton Portland Cement Company. Market gardeners John and Thomas Laverack gave a name to Laverack Close.

Holy Green is nothing religious but named after resident Thomas Holy. Horner Road is after Henry Horner, secretary of the Mount Pleasant Land Society. Jepson Road is after tool manufacturer Jepson Milner. Kaye Place has roads built by joiner and builder David Kaye. Kenyon Street is named after John Kenyon, owner of the merchants, filesmiths and sawmakers whose factory was on this site.

Holmshaw Drive remembers Robert Holmshaw, scissor grinder and founder of the Scissor Grinder's Union. Blacksmith John Hoyland left his mark on the map in the form of Hoyland Road.

Blacksmith George Heathcote gave his name to Heathcote Street. Heath Road took the name of joiner and builder Henry Heath. Grocer John Hessey gave his name to Hessey Street. Joiner and builder Thomas Hicks is commemorated by Hicks Lane and Hicks Street. Holdsworth Street is after Albert Holdsworth, a shareholder of the Sheffield Brick Co. Ltd. William Hinchcliffe had a beerhouse in what is now Hinchcliffe Walk.

Gregory Road recalls builder and contractor John Gregory. Former abbot of Beauchief Abbey William Gresley gave his name to Gresley Road. Both Haggard Road and Rider Road are named after author Sir Henry Rider Haggard. Hands Road recalls silversmith Thomas Hand. Hanson Road remembers the Rev John Hanson, minister of Loxley Chapel. Hardcastle Drive was inspired by Hannah Hardcastle who ran the local dame school. Haslehurst Road was named after a partner in the Norfolk Brewery, Charles Haslehurst.

Forbes Road takes the name of local doctor Alexander Forbes. An official of the Freehold Land Society, Frederick Whitby is remembered by Frederick Street and Whitby Road. Froggatt Lane is after the merchant James Froggatt. Dr Alfred Gatty, vicar at Ecclesfield for no less than sixty-three years, editor and childrens' author, certainly justifies the naming of Gatty Road. Gifford Road is from Eric Gifford VC, 3rd Baron Gifford, whose military career saw action with the Ashanti expedition and the Zulu War. Glover Road recalls Sir John Glover, who accompanied Gifford on the Ashanti expedition.

Without the people there would be no history and even those who held no land and held no office are commemorated. Adsetts Street recalls George Adsetts, land agent for the land society. Aldred Road comes from Dr John C. Aldred, a medical officer. Local architect Gustave Alsing was the inspiration for Alsing Road. Local churchman Joseph Angus produced Angus Square. Builder William Anson gave his name to Anson Street.

President of the land society John Amstead gave his name to Amstead Road. The former Atlas Works stood in what is now Atlas Street. One of history's names is seen in Babington Crescent, this recalling Anthony Babington, the man whose name was given to the plot to remove Elizabeth I from the throne and replace her with Mary, Queen of Scots. Bardwell Road was named after auctioneer Thomas Bardwell.

Bland Street recalls John Bland, former landlord of the Brunswick Hotel. Blast Lane is the site of the blast furnace. Broughton Villa, home of businessman John Deakin, gave us Broughton Road. Caxton Lane comes from the name of the house of Joseph Pearce, founder of the *Sheffield Daily*

Telegraph who named his home after William Caxton, the most famous name in printing in this country.

Builder Reuben Clarke is remembered by Clarke Square, this is not the case with Clarke Street, which recalls another builder and joiner, John Clarke. Builder and developer Martin Cruise gave his name to Cruise Road. Dent Lane was home to the headquarters of the Dent Colliery. Dominoe Grove shares this mining background, this after Dominoe Joe who led a march protesting againt a reduction in wages. Medical man Joseph Turner came to Doctor Lane following his retirement through ill health.

Doncaster Street remembers steel man Daniel Doncaster. Emily Road recalls a relative of local estate-owner Joshua Spooner. Victualler Joseph Exley helped develop what is now Exley Avenue. Farish Place is after the first vicar of St Mary's, Rev Henry Farish. Similarly the name of the Rev Henry Favell, vicar of St George's, comes to Favell Road.

Place names from other parts of the land were once considered ideal sources for street names, as existing names they could not be considered in any way controversial. Examples include Alderney Road and Arras Street. Donetsk Way is also a place name, this inspired by the town twinning agreed between Donetsk and Sheffield. Old pub names are remembered in Angel Street, Ball Lane, Black Swan Walk, Cornish Street, and Crown Place. Some refuse to fit into any category, such as Adastral Avenue, from the motto of the RAF of *Per Ardua Ad Astra* or 'Through difficulties to the stars'.

Others include Acacia Road, one of a theme of flowers and shrubs. Painters and artists are seen in Chatfield Road and Crosby Road. Writers are another source, poet Ebenezer Elliott is remembered by Elliott Road, Meynell Crescent remembers poet Alice Meynell. Collingbourne Avenue is one example of the theme of forests, while lakes have given us, among others, Crummock Road.

Pub names begin with the Crown and Cushion, a variation on a popular name showing support for king and country by adding the cushion on which the crown is carried at the coronation. Both country and monarchy are mentioned in the Rose and Crown. Quite often such were named around the time a new monarch ascended to the throne. Nearly all pubs named the Prince of Wales were named during the Victorian era, hence this referred to the future Edward VII. In the Queens Head we see a reference to royalty in general, although often a specific queen is depicted by the sign-painter.

Perhaps the most common method of referring to royalty and the nobility is through heraldry. Hence the great number of pubs named after seemingly oddly-coloured animals, such as the Golden Lion, a powerful image favoured by many families including the Percy family, dukes of Northumberland, and King Henry I. The Rising Sun is another, this used by Edward III, Richard III and many other nobles. Edward IV, the dukes of Norfolk and the earls of March chose the White Lion for their coat of arms. In the Red Lion we find either John of Gaunt or Scotland.

The long relationship between the inn and the traveller and his mount is reflected in pub names and quite rightly as they were the motorway service station of the time. Hence the Horse and Groom came into being when the inn-keeper also offered stabling for a weary horse while the Travellers Rest focused on the rider. Another equine reference is found in the Saddle. The Waggon and Horses recalls the time when goods were all transported by road, until the advent of the canals. The same is true of both the Old Horse and Cart and the Packhorse Inn. The blacksmith would often set up shop next to pub, this being advertised in names such as the Anvil. His tools provided the inspiration for the name Hammer and Pincers.

If not advertising the services then promote the product, this was how the pub sign and thus the first names began. In the Three Tuns we see a reference to the large barrels holding wines and beers, with the number appearing in the arms of both the Worshipful Company of Vintners and the Worshipful Company of Brewers. A larger container proved to be the inspiration behind the Hogshead, this holding some 250 gallons, while the Barrel Inn requires no explanation and neither does the Tap and Barrel.

In the Grapes we see an obvious reference to the ingredient producing wines, and is perhaps the oldest pub name in the country as it was certainly used as an advertisement in Roman Britain. That of the Vine Inn is a different slant on the same name. The Punch Bowl is another and it may come as a surprise to discover punch is by no means a modern concept, the first mention in England dates from 1630.

Of course not only drink is found within but food, too, although it is rare to find this as a pub name for menus change so often. Pub games have been popular for centuries, sometimes as a specific sport as in the Anglers Rest, or Cricketers Arms, and other times as a general invitation at the Sportsman Inn.

Location is also important, such as in those visiting the Museum public house or the common name of the Railway and also the Station. The Navigation fits into this category as it is only ever found by navigable waterways, while every stop in the coaching era was a pub and the Coach and Horses advertises itself as the equivalent of a bus stop. The Bridge Inn names its own marker. A prominent marker was the original idea behind the number of trees found in pub names, although the added bonus of an attractive image would also have been a factor. Here we see the Elm Tree Inn as one example and the Oak Inn another.

For pubs in an area where many are employed in a certain trade then offering their services to same hopefully gives the impression it is there for them. In the Pen and Wig we see a name associated with the legal profession. The Bankers Draft may be a financial reference but is also a clever play on words referring to draught beers. The Graduate hopes to attract the trade of the student population.

The Firkin Brewery used alliteration and their own name in all their pubs. Normally the first element has no etymological value whatsoever but in the Steel City they came up with the Foundry and Firkin. This was not the case when they came up with the Floozy and Firkin. If Sheffield is famous for one thing it is the steel industry and in particular cutlery, hence the name of the Cutler almost suggested itself but not quite as readily as the Steel Inn. Another skilled job gave a name to the Masons Arms and also the Freemasons Arms, while many workers would have found themselves drawn to the Miners Arms.

One of the earliest and most common names is the Plough, the image offering refreshment to those who worked the land and at one time this was almost everyone. Following the use of the plough a second tool was dragged across the field to break up the clods of soil and make the land more manageable, that tool gave a name to the Old Harrow. The Nailmakers Arms will have been one of the first to come from the metal industries, as will the Foundry Arms.

Whilst religion could never be said to be a trade as such, it has provided many more pub names than we would expect. That of the Priory, the Angel and also the Old Monk are fairly obvious, however a little thought is required to see the biblical story of the flood behind the Dove and Rainbow. The rainbow was the sign given by God he would never send another flood of such magnitude upon the Earth again, the dove the first to return with evidence the waters were receding. The survivors and progenitors of all life on the land were aboard the vessel shown outside the Noah's Ark public house. The Lion and Lamb are symbols of the Resurrection and the Redeemer, respectively.

Famous figures commemorated by a pub name include the Bard of Avon, indeed only one has more named after him and includes the Shakespeare. The one with the most named after him is the hero of the Battle of Trafalgar and, as if to prove this, there are two differently named premises in Sheffield, the Nelson and the Lord Nelson. Often the George Inn is mistakenly thought to recall one of the six kings of this name and yet almost without exception this is a reference to St George, the patron saint of England. The Robin Hood Inn may depict a figure in Lincoln green but originally this was no outlaw of folklore, this was the important position of woodsman or forester. George Brydges, Lord Rodney, enjoyed a successful career in the Royal Navy rising to the rank which gave a name to the Admiral Rodney.

Not only people but events are commemorated by pub signs. One of the most easily recognised images will be the original adhesive stamp and probably the most famous of all, and today one of the most valuable for collectors, is shown on the sign outside the Penny Black. The Royal Oak recalls the story of the escape of Charles II and when he hid in the Boscobel Oak just above the heads of the Parliamentarians. Perhaps the Three Merry Lads also fits into this category. When this pub was run by the Marsden family in the early nineteenth century they had three sons who gave their name to the pub.

Silkstone

Records of this name include the Domesday listing of Silchestone. This is Old English, with the Saxon personal name followed by *tun* and describing 'the farmstead of a man called Sigelac'.

Bull Haw Hall took an existing name, one referring to 'the nook of land where bulls are reared'. Lindley is from *lind leah* or 'the woodland clearing where lime trees grow'. Nabs House is from Old Scandinavian *nabbi* and referring to 'the peak, or knoll'. Noblethorpe features the elements *nobil thorpe*, here the 'outlying farmstead' is highly unlikely to have belonged to nobility as the term is simply not used in this context, hence this was probably used as an early surname or a nickname. Icicle Wood has nothing to do with temperature, this comes from *ecels* and informs us this was 'land added to an existing estate'.

Coal was transported out of Silkstone via a waggon-way operating from the early nineteenth century. Today this is marked by a memorial in the village.

above: Silkstone advertises its past; *right*: Silkstone Common sign

Skelbrooke

Found as Scalebre in 1086, Scelebroch in 1160, Skelebroke in 1220, Skelbroke in 1298, and as Skeldebrok in 1315, this is derived from *scela broc* and speaks of 'the brook at, by or near a shieling'.

Locally we see Skelbrooke Collins, the addition from Old English *holegn* shwing 'holly' grew here. Again the name has an addition in Shelbrooke Rein is from *reinn* meaning 'boundary strip of land'. A similar reference has given rise to Tongue End, from *tunge* and describing 'the tongue of land'. Today the meaning behind Muckey Lane can still be seen, this from *muk hyll lane* it tells us this was 'a muddy route on the hill', probably down to the number of wheels and hooves passing along here churning up the surface.

Skellow

Found as Scanhalle in Domesday and as Scalehale in the late twelfth century, here is Old English *scela halh* or 'the nook of land with a shieling'.

Undoubtedly the Skell River has been influenced by the name of Skellow, and *vice versa*. However this tributary of the Old Ea Beck has a different origin, here Old Scandinavian *skjallr* means 'resounding', a reference to it being swift flowing and noisy.

The English Civil War came to Skellow, the Parliamentarians built mounds for cannon to cover the Great North Road, these mounds still exist and have given a name to the fifteenth century inn known as the Bridge House.

Skellow sign

Sprotborough

Recorded in Domesday as Sproteburg, this is an Old English place name with two possible meanings. Either this is *sprota burh* and 'the stronghold overgrown with shoots' or 'the stronghold of a man called Sprota' if the first element is a Saxon personal name.

Local names include Cusworth, where the suffix is Old English *worth* follows the personal name to speak of 'the enclosure of a man called Cuthsa'. The Ivanhoe is a local pub named after the eponymous character in Sir Walter Scott's novel published in the early nineteenth century.

Stainborough

Listed as Stanburg in 1086, as Stanburght in 1522, and as Steinburch in 1160, these records show this was originally Old English *stan burh* later to be replaced by the Old Scandinavian *steinn* but keeping the meaning of 'the stony fortified place'.

Locally we find Falthwaite, this from *falh thveit* and understood to refer to 'the clearing broken up for cultivation' or 'the clearing allowed to lie fallow'. Bagger Wood gets its name from Old English *baggere* 'a hawker'. From the same language comes Ivas Wood, *ifig hyrst* describing 'the wooded hill where ivy abounds'. Wentworth Castle takes the name of the man whose money built it, Thomas, Earl of Strafford and Wentworth. The local pub also remembers this family being known as the Strafford Arms.

Staincross

Domesday records this as Staincros, a fairly obvious reference to a 'stone cross' and coming from Old Scandinavian *steinn kros*. That stone cross will have been a marker, likely for the same reasons as this being the meeting place for the Staincross wapentake. A wapentake was an administrative area, the Scandinavian term and the equivalent of the Saxon hundred, itself self-explanatory as the region covering approximately one hundred farmsteads.

Whilst the Saxons and Scandinavians saw Staincross as an important location, not until the twenty-first century did the postal service recognise this as a village. Maps have never adequately defined the boundaries of Staincross with neighbouring Mapplewell and Darton. Villagers campaigned for the right to use Staincross as a part of their official address and a ballot in 2003 saw a vote overwhelmingly in favour of doing so.

Stainforth

The second 'stony ford' is in South Yorkshire. Again it comes from Old English *stan ford*, this name is recorded as Steinforde in Domesday.

Stainton

Listed as Stantun in the Domesday record of 1086, as Steintun in 1196, and as Stenetun in 1375, the name was originally Old English *stan tun*. However the later forms show the influence of the Norsemen when Old Scandinavian *stein* replaced *stan* without changing the meaning of 'the stony farmstead'.

Locally we see Wilsick Hall, this from *wilig sic* and speaking of 'the drain or small stream by the willow trees'.

Deepcar sign, near Stocksbridge

Stocksbridge

Whilst there is no record of this name before 1841, it seems this must represent Old English *stocc brycg* and describe 'the bridge made from logs'.

The Miners Arms, Stocksbridge

Sutton

Listed as Sutone in 1086 and as Suttona in 1180, this common place name is from Old English *sud tun* and speaks of 'the southern farmstead'.

Locally we see Wetflat Plant, a name featuring three elements derived from two time periods. The wood plantation was a much later addition to what was *wet flat* or 'the watery level meadow'. Just as simplistic is the name of Camp, this a pointer to an ancient earthwork.

Midhopestones sign, east of Stocksbridge

Swinton

One of two places named such in the traditional county of Yorkshire. Here is a name describing 'farmstead for pigs', this from Old English *swin tun*. The name is found as Suintone in the Domesday survey of 1086.

Pubs include the Butchers Arms, a reminder that ale would never be enough to bring a living wage to an inn-keeper and forced him to have a second string to his bow. With a cellar offering a cold environment, taking up butchery made perfect sense.

Sykehouse

A name listed as Sikehouse in 1404, this is of Old Scandinavian derivation in *sik hus* and describes 'the house by the stream'.

Canal junction at Sykehouse – photograph courtesy of Wikipedia Creative Commons

T

Tankersley

Documented as Tancresleia in Domesday, the eleventh century record pointing to a combination of an Old German personal name and Old English *leah* and speaking of 'the woodland clearing of a man called Tancred or Thancrad'.

Pilley comes from Old English *pil leah* and speaks of 'the woodland clearing where shafts are obtained'. This description enables us to paint a mental picture of life in Saxon England where the trees around the clearing had been pollarded to produce new growth in the form of straight poles, here they were trimmed and used for the shafts of spears and other implements. The local is the Gate Inn, a name which ultimately comes from Old English *gaet* or 'way'.

Thorne

Listed in Domesday as Torne, this name comes from Old English *thorn* meaning the '(place at) the slope'.

Here we find Inkle Moor, a name with few records but which probably represents 'the moorland of a man called Enkil'. Old Scandinavian *hegn* means 'enclosure' and is seen today as Hains Farm. New River is not only a tributary of the River Don it is the new course of that river, hence the name.

Pub names begin with the Punch Bowl, clearly an advertisement for the product. It may come as a surprise to find the drink is not a comparatively recent addition to the table at parties, the first record of punch is in an early seventeenth century document. There has always been some dispute as to the etymology of the name, some saying it is derived from the word for 'five' in several of the languages of the Indian sub-continent and this refers to the number of traditional ingredients. There is no record of there ever being just five ingredients in punch. Much easier to see is the location advertised by the Canal Tavern. In the Green Dragon we see an heraldic reference to the earls of Pembroke, important landowners. The Rising Sun is also heraldic, but so common in coats of arms it is virtually impossible to know the source.

There are two local features of note, including the parish church dedicated to St Nicholas. Most of the church dates from between the twelfth and fifteenth centuries, although there are later additions, and it has been given a Grade I listing. We also find Peel Hill Castle, there are no remains of the Norman building erected by the de Warenne family but the foundations reveal something of its construction. Records show it was connected with Hatfield Chase, meaning it was likely used as a hunting lodge, and the tower housed prisoners during the sixteenth century, a hundred years before it was demolished.

Thorpe Salvin

Another of the many examples of Old English *throp* or 'outlying farmstead', here the addition is a reminder of the Salvain family, lords of this manor from

Old Hall Thorpe Salvin – photograph courtesy of Wikipedia Creative Commons

the thirteenth century. Records of this name include Torp in 1086 and as Thorpe Saluayn in 1255. There is also a record of this place as Rynkenild Thorp, the earlier addition a reference to the nearby Roman road known as Rykneild Street.

These are not the only potential names when naming new streets. In the eleventh century this was a part of the manor of Roger de Busil. The lords of the manor later lived at the now demolished Thorpe Hall, first built by Robert Smythson in 1570, which had replaced an earlier manor house on this site and the new home of Brian Sandford. Within seventy years the hall had been bought by Edward Osborne, but by the time his descendant Thomas Osborne had inherited the estate and he had become Duke of Leeds, the hall was already falling into a state of disrepair.

Thrybergh

Recorded as Triberge in the Domesday record of 1086, here is a Scandinavian personal name and Old English *croft* telling of 'the small holding of a man called Thorir'.

Reresby Road recalls the family who were associated with the village for several generations. From the early fourteenth century we find one Sir Adam de Reresby. Fifty years later we find Sir Thomas de Reresby with his wife Alice and their children, also Thomas and Alice, and this pattern continued until the seventeenth century. Sir John Reresby, a favourite of both Charles II and James II, headed a manor realising an impressive £1,700 per annum at the time of his death in 1689. Yet before the end of the century his son and heir, Sir William Reresby, had gambled away every last penny.

Eventually the estate reached the Honourable John Finch, son of the Earl of Ailesford, and whose name is remembered by Finch Close. Fullerton Close recalls the maiden name of his wife Judith, daughter of John Fullerton. The family are also remembered by Fullerton Church of England School, built in 1819.

Thurgoland

Listed as Turgesland in Domesday, here is 'the cultivated land of a man called Thorgeirr', a Scandinavian name with *land* following the personal name.

Stubbin Wood has changed little since the Saxon term *stubbing* referred to a 'man-made clearing'. Even more obvious is Cheese Bottom from *cese botm* and indeed referring to 'the valley bottom producing good cheese'.

above: Green Dragon public house, Thurgoland; *right*: Thurgoland's parish clock

Thurlstone

A name combining an Old Scandinavian personal name with Old English *tun* and describing 'the farmstead of a man called Thorulfr'. This place appears as Turulfestune in Domesday.

From *bula hus* comes Bullhouse which means exactly what it seems, although perhaps 'stall or shed' would be more accurate than 'house'. Hazlehead was named for being 'the headland where hazels grow'. Carlecotes comes from *ceorl cot* 'the cottages of the peasants'. Ecklands is thought to represent 'the hollow of the oaks'. Reddishaw Knoll is raised ground where 'the reedy copse' is found, this from *hreodig sceaga cnoll.* The same element is seen in Smallshaw, where *smael sceaga* tells of 'the narrow copse'.

Softley comes from *softe* leah and refers to 'the pleasant or agreeable woodland clearing'. Hartcliff Hall comes from *heorot clif*, Old English for 'the bank where harts are seen'. Hornthwaite was 'the clearing at the horn of land'. Many boulders are still found at Ranah Stones, named from *hrafn haugr* and referring to 'the place of many stones'. Shore Hall also refers to the topography, *scora* meaning 'the steep slope'.

In Long Grain we find an origin of *lang grein* or 'the river fork'. Nubbuck End features a dialect *nubbock* meaning 'a lump' and used in a topographical sense. Harden speaks of 'the valley of the hares' from *hara denu*, while *gleoda nest cloh* speaks of 'the ravine of the nesting kites' and is today recorded as Glead Nest Clough.

Thurnscoe

While Domesday records this name as Ternusche we also find that of Thirnescoh in the late eleventh century. Here Old Scandinavian *thyrnir skegr* refers to 'the thorn tree wood'.

The Howell Beck flows through Thurnscoe before emptying into the Frickley Beck. This is from *hol bekkr* or 'the stream in a hollow'.

The former Roman road of the Ryknild Street ran near here, although its unofficial local name was Cow Track, a rather ignominious label given by those who knew it as being used to move dairy herds. The church of St Helen's has some remains from the original construction in 1087. During renovation work a skeleton was uncovered. Buried during the Saxon era, it shows this was a sacred site before the church was erected. It may also be evidence of how a new religion, such as Christianity, permanently replaced earlier pagan beliefs by utilising an existing sacred site simply by building a church on it and thus removing the former location where rituals were performed. Of course it is impossible to know just when Christianity became the dominant religion here and it may be the burial is an early Christian example.

Tickhill

Seen as Tikehill in a document dating from the twelfth century, here is 'the hill of a man called Tica' if this is a Saxon personal name with *hyll* or 'the hill where young goats are kept' if this is Old English *ticcen hyll.*

Castle Hill is a street name coming from the defensive feature built by Roger de Busli. The Norman castle has played a role in several of the most important periods in English history. It was a base for the supports of Prince John against the rightful king, his brother Richard I. A siege lasting some three weeks was an important episode in the barons' uprising of the early fourteenth century. During the English Civil War the Royalists used the castle as a local base of operations, resulting in it being disabled by Parliamentarians. Today the castle is a ruin, albeit an impressive one.

Dadsley Well is another street name, this the only surviving reminder of what was once 'the *leah* or woodland clearing of a man called Daeddi'. Minor places

St Mary's church, Tickhill – photograph courtesy of Wikipedia Creative Commons

names include Lindrick, which describes 'the strip of land where lime trees grow'. We also find Woolthwaite which, without Scandinavian influence, would have developed as Woolley. Both languages give the origin as 'the woodland clearing infested by wolves' and yet the use of 'infested' does not really work when speaking of mammals, only pests (ie: insects) infest a place and thus perhaps the first element has been corrupted so as to be no longer recognisable.

The Royal Oak is one of the three most common pub names in England and was unheard of before the middle of the seventeenth century, for it remembers one of the best known narratives in the history of our islands. The oak tree in question is the Boscobel Oak, a large specimen in Shifnal, Shropshire and where Charles II and his aide Colonel Carless hid from noon to nightfall to escape the Parliamentarian soldiers. Following his defeat at the Battle of Worcester in 1651, King Charles was aided by a number of his supporters as he attempted to flee the country. After a number of close shaves he was smuggled to France by members of the Pendrell family disguised as a woodman, complete with a severe haircut and stained face and arms to make it look as if he was used to working outside. Nine years later came the Restoration of the Monarchy, the 29th of May, the date of the king's birthday, being chosen as the official date and thereafter was known as Royal Oak Day.

The present tree is not the original, this was destroyed by souvenir hunters in the 17th and 18th centuries who removed pieces of bark and even whole branches. A second tree, from an acorn of the first, known as the Son of the Royal Oak grew until it was badly damaged by a storm in 2000. A third tree was planted as a sapling by Prince Charles in 2001, this had been grown from an acorn taken from the 'Son' and is thus 'the grandson'. Modern-day souvenir hunters can still acquire a piece of the history, for other grandchildren of the Boscobel Oak are on sale here. Buyers receive their small oak along with a certificate of authenticity.

Tinsley

Found as Tineslauue in 1086, the Domesday record pointing to this coming from a Saxon personal name and Old English *hlaw* and telling of 'the mound of a man called Tynni'.

The church of St Laurence may only date from 1877 but is on the site of a place associated with Christian worship since Saxon times. During the interim years an annual payment was made by the Crown to pay for a service for the dead of the village. A second unusual tradition concerned the Lord of Tickhill and the lord of the neighbouring manor of Tinsley. The latter was required to bring a pair of white gloves each Michaelmas, in return he would receive one white dove which he would keep over winter. Just what this exchange symbolises is unclear.

Todwick

A name combining a Saxon personal name and Old English *wic* and tells of 'the specialised farm of a man called Tata'. Listed as Tatewic in the Domesday survey, the speciality in these cases is invariably dairy produce.

Wasteneys Road can be traced to the family who held this manor. By 1303 Sir Edmund Wasteneys held this manor and, although he was then resident at Hedon in the Clay, there are records of this family in Todwick until 1743. This eighteenth century record shows they were no longer the affluent landowners but two families living in a cottage on waste land. The last recorded Wasteneys in Todwick was Thomas with his death in 1902.

Totley

A name combining a Saxon personal name and Old English *inga ham*, this is 'the homestead of the family or followers of a man called Tota'. This name appears in Domesday as Totingelei. The name has also been transferred to the Totley Tunnel, which takes the railway under Totley Moor and is the longest tunnel wholly within the United Kingdom.

The Fleur de Lys is heraldic, this French for 'lily flower' and used by numerous families, towns, cities and groups. It is thought the image was originally that of an iris, although some sources suggest a sceptre or an axe. The image is used by so many it is impossible to know the specific source but all are linked by France.

Treeton

Seen as Tretone in the Domesday survey, this is probably from the Old English *treow tun* and tells of 'the farmstead built with posts'.

In Ulley we see a local name from Old English *owl leah* or 'the woodland clearing frequented by owls'. Stoket Lane tells those who understand the origins it was 'the road to the stone bridge'. Catcliffe Flash is a fairly modern feature, a result of subsidence seeing the land alongside the Rother dropping and flooding to produce a lake and marshland now designated a nature reserve. Ironically the name of Catcliffe, which had existed for centuries before the region was mined, describes 'the raised bank of land of a man called Catta'.

Trent (River)

An ancient river name which can only be defined by association with other early tongues of the Proto-Indo-European group. This speaks of 'the trespasser' and a warning this river is prone to flood.

The Trent unites with the Ouse to form the Humber. Despite this confluence being known as Trent Falls there is nothing resembling a waterfall, the name refers to the race of water created by the tides. It is commonplace for boats to have to await the turn of the tide in order to manoeuvre from the Trent to the Ouse and vice versa. What is less well known is the Trent, like the Severn, has a tidal bore. Known as the Trent Aegir, said to be named for the Norse god of the sea, it reaches a height of around five feet. However it peters out at Gainsborough owing to the shape of the river bed, what little visibly remains cannot negotiate the weirs encountered soon after.

The River Torne is a tributary of the Trent, this a far more recent name where back-formation from the Torne Bridge has clearly been misunderstood to have been named from the river. Yet we do know where the bridge name comes from a lost place name, that of Torne Wath a minor name in neighbouring Rossington and originated as 'the ford of the thorn tree'.

U

Ulley

Found in the pages of Domesday as Ollei, this is from Old English *ule leah* and describes 'the woodland clearing frequented by owls'.

During the Second World War bombers heading for Sheffield dropped their payload on the Main Street through Ulley. The bombs failed to detonate, thus the cottages were evacuated and the bomb disposal team sent for. After assessing the situation, a meeting was convened to decide on a course of action, hence everyone attended the meeting place – the local pub. Whilst discussing their plans the bombs exploded, utterly destroying the cottages.

W

Wadworth

From a Saxon personal name and Old English *worth*, this describes 'the enclosure of a man called Wada'. The name is recorded as Wadewrde in Domesday.

Minor Old English names of note include Stancill, a name from *stan sael* describing 'the stone dwelling hall', and Wellingley, which speaks of itself as 'the dwellers by the stream'.

Pubs include the Fox and Hounds, a reminder that hunting has been a major event in the history of the countryside and the pub will have played a central role as both a meeting place for the hunt and where the kennels were often located. We also find the White Hart which, as with so many coloured animal names, began as an heraldic sign. Yet the short reign of Richard II is not the reason for the popularity of this name, it is entirely down to this becoming the generic name for every pub for centuries afterwards.

Winding Wheel in Wales – photograph courtesy of Wikipedia Creative Commons

Wales

A name found in the Domesday survey exactly as it appears today. As with the principality of this name, here is Old English *walas* and refers to the '(place of) the Britons'. While the definition is given here as 'Britons', it seems likely the Saxons used this to describe any culture or nationality other than their own, thus should probably be seen as meaning 'foreigners'.

The minor name of Kiveton features the common Old English *tun* with a Saxon personal name to describe 'the farmstead of a man called Cifa'.

Warmsworth

Seen as Wermesford in Domesday and as Wermesworth in the early twelfth century, this is derived from a Saxon personal name and Old English *worth* and tells of 'the enclosure of a man called Waermi or Waermund'.

Two local street names can be traced to one individual from the family who were lords of the manor of Warmsworth. Cecil Avenue and Wrightson Avenue were inspired by the most impressively named Lady Isabella Georgiana Katherine Battie-Wrightson, nee Cecil, third daughter of the Marquis of Exeter of Burghley House.

Wath upon Dearne

Another place from Old Scandinavian *vath* in a plural form and describing 'the fords', this place is recorded in Domesday as Wade. Here the addition is a river name discussed under its own entry.

Here we find the hamlet of Brampton Bierlow, two names which come from two different languages. The common English name of Brampton is derived from Old English *brom tun* or 'the farmstead where broom trees grow'. Normally such common names are found with an addition, either comparative such as High, Great, North, East, Dry, etc., or an addition from the family who were lords of the manor, most often this would be the Norman landholder. Here the name comes from old Scandinavian, where *byjar log* literally described 'the law of the village' and effectively told us it was considered a township in its own right. That there are two names from the same era is probably an indication both names were in use for the place.

Minor names include those named for associations with a person or group of people. For example Rainborough began as 'the hill of a man called Regen'; while Alderthwaite speaks of itself as 'the woodland clearing of a man called

Hoober Stand at Brampton Brierlow near Wath upon Dearne – photograph courtesy of Wikipedia Creative Commons

Alfward'; Wentworth was named as 'the enclosure of a man called Wintra'; Cortworth 'the enclosure of a man called Corta'; with Abdy, this pointing out it 'belonged to the abbey'.

Coaley Lane was originally known for running alongside 'the cold or exposed mound or tumulus'. Hoober had a very obvious feature in its 'hill with a jutting spur of land'. What was known as 'the enclosure where flax was grown' is now known as Linthwaite. Royds is from *rod* in a plural form referring to 'the clearings'. Upper Hoyland and Nether Hoyland share a name from *hoh land* 'the agricultural land at the hill spur'. Skier's Hall derives its name from the Danish or Norwegian word *skur* meaning 'a shed', which would have been here centuries before the name was taken for the hall. Swinton is a fairly common name and thus we would expect a fairly simple meaning, finding such in 'the farmstead where swine are kept'.

A quite fitting origin is found in the roads which now occupy the former route of the Dearne and Dove Canal. Opened between 1798 and 1804 it served the local collieries of the Dearne Valley. Other goods were transported, including lime, glass, oil products and pig iron. The canal went into decline following the arrival of the railway and subsidence created by the very coal mines it was cut to service. One particularly wide section was nicknamed the Bay of Biscay by locals and the road which now follows this route is now officially known as Biscay Way.

For the author there is no more satisfying derivation than those which enable a mental picture to be drawn of the place in the days when the name was first coined or, as in the case of Stubbin, shortly before the name was first seen. This name describes 'a place where tree stumps have been cleared', and a simple task to image men labouring in the fields to remove the stumps which would otherwise be most troublesome to the ploughman.

Pubs include the Cottage of Content, an invitation suggesting this as exactly what a customer would find within. At the Dearnesman the invitation is more specific, inviting the locals. The Church House shows this stands on church land, while the Saracens Head is heraldic, this image used by families whose ancestors journeyed to the Holy Land on one of the Crusades.

Wentworth

Listed as Wintreuuorde in 1086, the Domesday record describes 'the enclosure of a man called Wintra' with the Saxon personal name suffixed by Old English *worth*.

By far the most imposing building in the village is Wentworth Woodhouse, a country house boasting no less than twenty-one follies including the Bean Seat, the Doric Temple, the Ionic Temple, the Mausoleum, Needle's Eye, and the Vinegar Stone. We also find Keppel's Column, built in the late eighteenth century to mark the acquittal of Admiral Augustus Keppel, who had faced a court-martial after the Battle of Ushant, having been accused of deliberately running his vessel *Maidstone* aground when pursuing a French vessel. An internal spiral staircase allows access to the top of the 115 feet high tower, itself sitting atop a hill rising to 480 feet above sea level. This makes it the highest point in the vicinity with the exception of Hoober Stand, another folly on the estate of approximately 100 feet high sitting atop a ridge of land rising to 515 feet above sea level. It was constructed to mark the role of Thomas Watson-Wentworth, Earl of Malton in quashing the Jacobite Rebellion of 1745.

Wentworth Castle Rotunda, Stainborough – photograph courtesy of Wikipedia Creative Commons

Wharncliffe Side

Listed as Querncliffe in 1406, the basic name is derived from Old English *cweorn clif* or 'the cliff where querns or millstones are obtained'. By 1634 this had become Wharnetcliffe Side, the addition of Middle English *side* further emphasising its location on 'the hill side'.

Nearby Wharncliffe Crags was the site of the slaying of the dragon by a knight known as Moore of Moore Hall (now seen as More Hall). The tale tells of how he procured a suit of armour bristling with metal spikes. In what cannot be called a chivalrous act he approached the fearsome beast and kicked it as hard as he could in its only vulnerable spot. Such a kick in the dragon's "arse-gut" proved fatal, the whole scene portrayed in the stonework of Sheffield Town Hall.

Wharncliffe Side sign

Whiston

Recorded in Domesday as Witestan, this is Old English *hwit stan* meaning simply 'the white stone'.

Also seen as a district name, Morthen is generally accepted as *mor thing* 'the moor of the assembly', a meeting place. However when applied to the hamlet rather than the village there is a local tradition of this being from a combination of Old English *morth* and Old Scandinavian *eng* and giving 'the meadow of the murder'. Old Scandinavian *gil thveit* has given us 'the woodland clearing in a ravine', which appears on modern maps as Guilthwaite. In Blue Man's Bower we have a later name which is derived from Middle English *blewman bur*. This satisfies the author's quest for place names giving a snapshot of life in earlier times, although the *bur* or 'cottage' was not the home of a blue man but a black man.

Locals enjoy a drink at the Chequers. This is probably one of the oldest pub names in England, it was certainly used by the Romans for their inn-keepers showed this to tell there was a board game being played within. Later, when it was adopted as the sign showing the inn inn-keeper was also a moneyer. Today we still use that reference in the title of the office charged with seeing to the nation's purse strings, the Chancellor of the Exchequer. Another sign gave a name to the Golden Ball, this bright image has been adopted by many traders.

Whitford

Had this been of Old English derivation this would have been easy to see as 'the white or bright ford'. However the origin is not Old English but Old

Scandinavian and the meaning is quite different. From a personal name suffixed by *gipt* and listed as Witegift around 1070, this is 'the dowry land of a man called Hvitr'.

Wickersley

Found as Wicresleia in the Domesday record of 1086, this is a combination of a Scandinavian personal name and Old English *leah* and describing 'the woodland clearing of a man called Vikarr'. The manor was once held by the family who adopted the surname of de Wickersley, seemingly preferring it to his traditional Norman surname of FitzTurgis.

Wombwell

A name with two possible origins which is recorded as Wanbuelle in Domesday. This is either Old English *wamb wella* and 'the spring in a hollow' or 'the spring of a man called Wamba' if the first element is a Saxon personal name.

Locally we see the Knoll Beck, a name derived from the *cnoll* or 'hillock'. The River Dove, a tributary of the Dearne, comes from the British or Celtic *dubro* and describes 'the dark one'. When it comes to pubs here we find a reference to product in the Barley Sheaf, this appearing on the arms of the Worshipful Company of Brewers. The George is a patriotic establishment, not a reference to the six kings but to the patron saint of England. When it comes to patriotism the efforts of the leader of the Peasants' Revolt in 1381 are marked by the Wat Tyler Inn. His efforts were on the point of succeeding, Richard II had agreed to many demands, when he was killed by the Lord Mayor of London and his supporters crushed by military force.

Woodhouse

Recorded as Wdehus in a documented dated 1200, this is derived from Old English *wudu hus* and tells of 'the house in or by the wood'.

Woodhouse Doric Lodge at Wentworth – photograph courtesy of Wikipedia Creative Commons

Woodsetts

Seen as Wodesete in 1324, here is 'the animal folds in the wood' from Old English *wudu set*.

Hoades Farm has given us Hoades Avenue, this recorded as Audzus on the first map of the village dated 1771.

Worrall

Documented as Wihale in 1086 and as Wirhal in 1218, this is 'the nook of land where bog myrtle grows' from Old English *wir halh*.

Worsbrough

Listed as Wircesburg in 1086, the Domesday listing shows this to be 'the stronghold of a man called Wyrc' where the Saxon personal name is followed by Old English *burh*.

Here are several places named for a woodland clearing, some featuring the Old English *leah* others Old Scandinavian *thveit*. Darley Hall and Darley Cliff find *deor* giving 'the woodland clearing frequented by deer'. Similarly Rockley Abbey features the element *hroc* to give 'the woodland clearing frequented by rooks'. When Ouslethwaite Hall was built it took the existing place name to describe the *osle* or 'blackbird woodland clearing'. A similar feature but man-made is seen in Wigfield. Here the Old Scandinavian suffix *fall* or 'place where trees have been felled' follows *vigg* or 'steed'. However if the first element is Old English, then this is *wigga* meaning 'beetle'.

Topographical features have given names to several places including Blacker, this from *blaec kjarr* or 'the black marsh'. Middle English *swathe* has given us Swaithe or 'the strip of grassland'. Balk comes from *balca* or 'the boundary ridge'. Friar Trail has no connection to the church or to a path, this from *frere taegl* or 'the projecting piece of land'. Lewden began as *hleow denu* 'the sheltered sunny valley'. Lockeaflash has retained much of the original form of *locere flasshe* to refer to 'the swamp associated with the shepherd'.

In Lobwood Lane we see a description of the route which ran alongside or through a wood. That description depends upon whether this comes from Old English *lobb* and referring to the branches as 'low hanging' or, should this be Old Scandinavian, the very similar *lobbe* warning of 'spiders'. Daw Croft recalls this small-holding was known for its 'jackdaws'. Finally influential families left their names in the landscape with Edmunds Row and Marrow House, both surnames reproduced perfectly as place names.

The Edmund family are also seen in the Edmund Arms, the place where locals can enjoy a glass of their favourite tipple. The Yews is one of the many pubs which utilise trees as a marker, these long-lived trees also made for an attractive image on the sign. The Dalesman could be seen as a patriotic sign, albeit a local one, while also offering an invitation to those in the locality as is the case with the Darley Inn. The Greyhound Inn features an heraldic image, one representing the dukes of Newcastle. Trades are an important source and no explanation is required for the Button Mill Inn.

Wortley

A name which tells of 'the woodland clearing where vegetables were grown' is recorded in Domesday as Wirteleie and is derived from Old English *wyrt leah*.

Bromley is a common enough place name in England. Whether a parish name or, as here, a minor one the origin is from *brom leah* 'the woodland clearing where broom grows'. Howbrook comes from *hol broc* 'the brook in a hollow'. From *fenkel straet* comes Finkle Street and describing 'the road with a bend in it'. Cob Castles features the element *cobbe*, here the 'lump' was topped by a large building rather than the castle we would picture.

Wortley Hall – photograph courtesy of Wikipedia Creative Commons

Wharncliffe comes from *cweorn clif*, a reminder this was 'the bank where millstones are obtained'.

Hobb Stones was probably the location for some local folklore, for *hob* points to 'the stones associated with the goblin'. Hermit Hill does indeed refer to the location of a hermitage. While the *thorp* or 'outlying farmstead' of Shipmanthorpe is clear, this is not the case with the reamining elements. Either this represents *schepman* or 'shepherd' or *scip man* quite literally 'ship man' or 'sailor'.

Wortley's most famous former resident is John Nevison. More often known as the seventeenth century highwayman Swift Nick, he is best-remembered for something which is more often attributed to Dick Turpin. Born here in 1639, he was the highwayman who made the famous ride from London to York to provide an alibi for a robbery of which he was undoubtedly guilty.

Y

Yorkshire

The county's name takes that of its historic heart of York and adds Old English *scir* or 'district', a Saxon administrative region and first referred to as such in the eleventh century. The city of York gets its name from an ancient Celtic tongue, either referring to 'the estate of a man called Eburos' or, and more likely, 'the estate associated with yew trees'.

Common place-name elements

Element	Origin	Meaning
ac	Old English	oak tree
banke	Old Scandinavian	bank, hill slope
bearu	Old English	grove, wood
bekkr	Old Scandinavian	stream
berg	Old Scandinavian	hill
birce	Old English	birch tree
brad	Old English	broad
broc	Old English	brook, stream
brycg	Old English	bridge
burh	Old English	fortified place
burna	Old English	stream
by	Old Scandinavian	farmstead
ceap	Old English	market
ceaster	Old English	Roman stronghold
cirice	Old English	church
clif	Old English	cliff, slope
cocc	Old English	woodcock
cot	Old English	cottage
cumb	Old English	valley
cweorn	Old English	queorn
cyning	Old English	king
dael	Old English	valley
dalr	Old Scandinavian	valley
denu	Old English	valley
draeg	Old English	portage
dun	Old English	hill
ea	Old English	river

east	Old English	east
ecg	Old English	edge
eg	Old English	island
eorl	Old English	nobleman
eowestre	Old English	fold for sheep
fald	Old English	animal enclosure
feld	Old English	open land
ford	Old English	river crossing
ful	Old English	foul, dirty
geard	Old English	yard
geat	Old English	gap, pass
haeg	Old English	enclosure
haeth	Old English	heath
haga	Old English	hedged enclosure
halh	Old English	nook of land
ham	Old English	homestead
hamm	Old English	river meadow
heah	Old English	high, chief
hlaw	Old English	tumulus, mound
hoh	Old English	hill spur
hop	Old English	enclosed valley
hrycg	Old English	ridge
hwaete	Old English	wheat
hwit	Old English	white
hyll	Old English	hill
lacu	Old English	stream, water course
lang	Old English	long
langr	Old Scandinavian	long
leah	Old English	woodland clearing
lytel	Old English	little
meos	Old English	moss
mere	Old English	lake
middel	Old English	middle
mor	Old English	moorland

myln	Old English	mill
niwe	Old English	new
north	Old English	north
ofer	Old English	bank, ridge
pol	Old English	pool, pond
preost	Old English	priest
ruh	Old English	rough
salh	Old English	willow
sceaga	Old English	small wood, copse
sceap	Old English	sheep
stan	Old English	stone, boundary stone
steinn	Old Scandinavian	stone, boundary stone
stapol	Old English	post, pillar
stoc	Old English	secondary or special settlement
stocc	Old English	stump, log
stow	Old English	assembly or holy place
straet	Old English	Roman road
suth	Old English	south
thorp	Old Scandinavian	outlying farmstead
treow	Old English	tree, post
tun	Old English	farmstead
wald	Old English	woodland, forest
wella	Old English	spring, stream
west	Old English	west
wic	Old English	specialised, usually dairy farm
withig	Old English	willow tree
worth	Old English	an enclosure
wudu	Old English	wood

Bibliography

Drinkall, Margaret and Drinkall, Chris *Rotherham Then & Now*
Dunkling, Leslie and Wright, Gordon *A Dictionary of Pub Names*
Ekwall, Eilert *The Concise Oxford Dictionary of English Place-Names*
Fisher, John *Anston in the County of York*
Harvey, Peter *Street Names of Sheffield*
Hoyle, E *Old Barnsley Streets and By-Ways*
Leslie, Jack and John *Down Our Streets*
Mills, A.D. *Oxford Dictionary of English Place Names*
Morris, R. W. *Yorkshire Through Place Names*
Munford, Anthony P. *A History of Rotherham*
Munford, Anthony P. *A Walk Around Rotherham: the Town Centre*
Newcastle upon Tyne City Libraries and Arts *What's In A Name?*
Thurlow, William *Yorkshire Place Names*

Doncaster Star
A Patchwork of Parishes